Communications
in Computer and Information Science   1620

More information about this series at https://link.springer.com/bookseries/7899

Alexei Pozanenko · Sergey Stupnikov ·
Bernhard Thalheim · Eva Mendez ·
Nadezhda Kiselyova (Eds.)

Data Analytics and Management in Data Intensive Domains

23rd International Conference, DAMDID/RCDL 2021
Moscow, Russia, October 26–29, 2021
Revised Selected Papers

Springer

Editors
Alexei Pozanenko
Space Research Institute of the Russian
Academy of Sciences
Moscow, Russia

National Research University Higher School
of Economics
Moscow, Russia

Bernhard Thalheim
Christian-Albrecht University of Kiel
Kiel, Germany

Nadezhda Kiselyova
A. A. Baikov Institute of Metallurgy
and Materials Science of RAS (IMET RAS)
Moscow, Russia

Sergey Stupnikov
Federal Research Center "Computer Science
and Control" of RAS
Moscow, Russia

Eva Mendez
Universidad Carlos III de Madrid
Getafe, Spain

ISSN 1865-0929   ISSN 1865-0937 (electronic)
Communications in Computer and Information Science
ISBN 978-3-031-12284-2   ISBN 978-3-031-12285-9 (eBook)
https://doi.org/10.1007/978-3-031-12285-9

This Springer imprint is published by the registered company Springer Nature Switzerland AG
The registered company address is: Gewerbestrasse 11, 6330 Cham, Switzerland

Preface

This CCIS volume published by Springer contains the proceedings of the XXIII International Conference on Data Analytics and Management in Data Intensive Domains (DAMDID/RCDL 2021) that was set to be held at the National University of Science and Technology MISIS, Moscow, Russia during October 26–29, 2021. However, because of the worldwide COVID-19 crisis, DAMDID/RCDL 2021 had to take place online.

DAMDID is a multidisciplinary forum of researchers and practitioners from various domains of science and research promoting cooperation and exchange of ideas in the area of data analysis and management in domains driven by data-intensive research. Approaches to data analysis and management being developed in specific data-intensive domains (DID) of X-informatics (such as X = astro, bio, chemo, geo, medical, neuro, physics, chemistry, material science, social science, etc.), as well as in various other branches of informatics, industry, new technologies, finance, and business, contribute to the conference content.

Previous DAMDID/RCDL conferences were held in St. Petersburg (1999, 2003), Protvino (2000), Petrozavodsk (2001, 2009), Dubna (2002, 2008, 2014), Pushchino (2004), Yaroslavl (2005, 2013), Pereslavl (2007, 2012), Kazan (2010, 2019), Voronezh (2011, 2020), Obninsk (2016), and Moscow (2017, 2018).

The program of DAMDID/RCDL 2021 was oriented towards data science and data-intensive analytics as well as data management topics. The program of this year included three keynotes. The keynote by Yibin Xu (Deputy Director of the Research and Services Division of Materials Data and Integrated System and leader of the Data-Driven Inorganic Materials Research Group at the National Institute for Materials Science, Japan) was devoted to the construction of an integrated materials data system for data-driven materials research. Emille E. O. Ishida (CNRS, Laboratoire de Physique de Clermont, Université Clermont-Auvergne, France) gave a talk on supervised (and especially active) and unsupervised machine learning and their application in astronomy for classification problems and the search for scientifically interesting anomalies. The keynote by Andrew Turpin (Associate Director of the Melbourne Connect and Director of the Melbourne Data Analytics Platform) discussed the development of a workforce of data and computer scientists that can support researchers at university to make use of digital technology in their research.

The workshop on Data and Computation for Materials Science and Innovation (DACOMSIN) constituted the first day of the conference on October 26. The workshop aimed to address the communication gap across communities in the domains of materials data infrastructures, materials data analysis, and materials in silico experiments. The workshop brought together professionals from across research and innovation to share their experience and perspectives of using information technology and computer science for materials data management, analysis, and simulation.

The conference Program Committee, comprised of members from 12 countries, reviewed 63 submissions. In total, 37 submissions were accepted as full papers and 15 as short papers and posters.

According to the conference and workshops program, 58 oral presentations were grouped into 13 sessions. Most of the presentations were dedicated to the results of research conducted in organizations located in Russia, including Kazan, Moscow, Novosibirsk, Obninsk, Tomsk, Tula, St. Petersburg, Petrozavodsk, and Voronezh. However, the conference also featured talks prepared by foreign researchers from countries such as Australia, Armenia, China, Finland, France, Germany, Japan, Italy, Sweden, and the UK.

For the CCIS conference proceedings, 16 peer-reviewed papers have been selected by the Program Committee (an acceptance rate of 25%), which are structured into four sections: Problem Solving Infrastructures, Experiment Organization, and Machine Learning Applications (three papers); Data Analysis in Astronomy (five papers); Data Analysis in Material and Earth Sciences (four papers); and Information Extraction from Text (four papers).

We are grateful to the Program Committee members, for reviewing the submissions and selecting the papers for presentation, to the authors of the submissions, and to the host organizers from the National University of Science and Technology MISIS. We are also grateful for the use of the Conference Management Toolkit (CMT) sponsored by Microsoft Research, which provided great support during various phases of the paper submission and reviewing process. The Organizing Committee wants to gratefully acknowledge the sponsor of the conference, Thermo-Calc Software AB, for their generous support. Thermo-Calc Software's mission is to develop computational tools that allow engineers to generate the materials data they need in their daily decision making to drive innovation and improve product performance.

June 2022

Alexei Pozanenko
Sergey Stupnikov
Bernhard Thalheim
Eva Mendez
Nadezhda Kiselyova

Organization

Program Committee Co-chairs

Alexei Pozanenko — Space Research Institute, RAS, and National Research University Higher School of Economics, Russia

Eva Mendez — Universidad Carlos III de Madrid, Spain

Program Committee Deputy Chair

Sergey Stupnikov — Federal Research Center "Computer Science and Control" of RAS, Russia

DACOMSIN Workshop Co-chairs

Nadezhda Kiselyova — Baikov Institute of Metallurgy and Materials Science, RAS, Russia

Vasily Bunakov — Science and Technology Facilities Council, UK

Alexandra Khvan — National University of Science and Technology MISIS, Russia

Organizing Committee Co-chairs

Mikhail Filonov — National University of Science and Technology MISIS, Russia

Victor Zakharov — Federal Research Center "Computer Science and Control" of RAS, Russia

Organizing Committee

Igor Abrikosov — National University of Science and Technology MISIS, Russia

Alexandra Khvan — National University of Science and Technology MISIS, Russia

Marina Nezhurina — National University of Science and Technology MISIS, Russia

Alex Kondratiev — National University of Science and Technology MISIS, Russia

Nikolay Skvortsov — Federal Research Center "Computer Science and Control" of RAS, Russia

| Vladimir Cheverikin | National University of Science and Technology MISIS, Russia |
| Irina Bajenova | National University of Science and Technology MISIS, Russia |

Coordinating Committee

Igor Sokolov (Co-chair)	Federal Research Center "Computer Science and Control" of RAS, Russia
Nikolay Kolchanov (Co-chair)	Institute of Cytology and Genetics, SB RAS, Novosibirsk, Russia
Sergey Stupnikov (Deputy Chair)	Federal Research Center "Computer Science and Control" of RAS, Russia
Arkady Avramenko	Pushchino Radio Astronomy Observatory, RAS, Russia
Pavel Braslavsky	Ural Federal University and SKB Kontur, Russia
Vasily Bunakov	Science and Technology Facilities Council, UK
Alexander Elizarov	Kazan (Volga Region) Federal University, Russia
Alexander Fazliev	Institute of Atmospheric Optics, SB RAS, Russia
Alexei Klimentov	Brookhaven National Laboratory, USA
Mikhail Kogalovsky	Market Economy Institute, RAS, Russia
Vladimir Korenkov	Joint Institute for Nuclear Research, Russia
Sergey Kuznetsov	Institute for System Programming, RAS, Russia
Vladimir Litvine	Evogh Inc., USA
Archil Maysuradze	Moscow State University, Russia
Oleg Malkov	Institute of Astronomy, RAS, Russia
Alexander Marchuk	Institute of Informatics Systems, SB RAS, Russia
Igor Nekrestjanov	Verizon Corporation, USA
Boris Novikov	St. Petersburg State University, Russia
Nikolay Podkolodny	Institute of Cytology and Genetics, SB RAS, Russia
Aleksey Pozanenko	Space Research Institute, RAS, Russia
Vladimir Serebryakov	Dorodnicyn Computing Center of RAS, Russia
Yury Smetanin	Russian Foundation for Basic Research, Moscow
Vladimir Smirnov	Yaroslavl State University, Russia
Bernhard Thalheim	Kiel University, Germany
Konstantin Vorontsov	Moscow State University, Russia
ViacheslavWolfengagen	National Research Nuclear University "MEPhI", Russia
Victor Zakharov	Federal Research Center "Computer Science and Control" of RAS, Russia

Program Committee

Alexander Afanasyev	Institute for Information Transmission Problems, RAS, Russia
Ladjel Bellatreche	National Engineering School for Mechanics and Aerotechnics, France
Dmitry Borisenkov	RELEX Group, Russia
Pavel Braslavski	Ural Federal University, Russia
Vasily Bunakov	Science and Technology Facilities Council, UK
Kheeran Dharmawardena	Cytrax Consulting, Australia
Boris Dobrov	Lomonosov Moscow State University, Russia
Alexander Elizarov	Kazan Federal University, Russia
Alexander Fazliev	Institute of Atmospheric Optics, SB RAS, Russia
Evgeny Gordov	Institute of Monitoring of Climatic and Ecological Systems SB RAS, Russia
Valeriya Gribova	Institute of Automation and Control Processes, FEBRAS, Far Eastern Federal University, Russia
Maxim Gubin	Google Inc., USA
Sergio Ilarri	University of Zaragoza, Spain
Mirjana Ivanovic	University of Novi Sad, Serbia
Jeyhun Karimov	Huawei Research Center, Germany
Vitaliy Kim	National Research University Higher School of Economics, Russia, and Fesenkov Astrophysical Institute, Kazakhstan
Nadezhda Kiselyova	Institute of Metallurgy and Materials Science, RAS, Russia
Alexei Klimentov	Brookhaven National Laboratory, USA
Sergey Kuznetsov	Institute for System Programming, RAS, Russia
Giuseppe Longo	University of Naples Federico II, Italy
Evgeny Lipachev	Kazan Federal University, Russia
Natalia Loukachevitch	Lomonosov Moscow State University, Russia
Ivan Lukovic	University of Belgrade, Serbia
Oleg Malkov	Institute of Astronomy, RAS, Russia
Sergey Makhortov	Voronezh State University, Russia
Yannis Manolopoulos	Aristotle University of Thessaloniki, Greece
Archil Maysuradze	Lomonosov Moscow State University, Russia
Manuel Mazzara	Innopolis University, Russia
Mikhail Melnikov	Institute of Molecular Biology and Biophysics, SB RAS, Russia
Alexey Mitsyuk	National Research University Higher School of Economics, Russia
Alexander Moskvitin	Special Astrophysical Observatory, RAS, Russia

Xenia Naidenova	Kirov Military Medical Academy, Russia
Dmitry Namiot	Lomonosov Moscow State University, Russia
Boris Novikov	National Research University Higher School of Economics, Russia
Panos Pardalos	University of Florida, USA
Jaroslav Pokorny	Charles University in Prague, Czech Republic
Natalya Ponomareva	Research Center of Neurology, Russia
Roman Samarev	Bauman Moscow State Technical University, Russia
Vladimir Serebryakov	Dorodnicyn Computing Centre of RAS, Russia
Nikolay Skvortsov	Federal Research Center "Computer Science and Control" of RAS, Russia,
Manfred Sneps-Sneppe	Ventspils University of Applied Sciences, Latvia
Valery Sokolov	Yaroslavl State University, Russia
Kirill Sokolovsky	Michigan State University, USA
Alexander Sychev	Voronezh State University, Russia
Bernhard Thalheim	University of Kiel, Germany
Alexey Ushakov	University of California, Santa Barbara, USA
Pavel Velikhov	TigerGraph, Russia
Alina Volnova	Space Research Institute, RAS, Russia
Alexey Vovchenko	Federal Research Center "Computer Science and Control" of RAS, Russia
Vladimir Zadorozhny	University of Pittsburgh, USA
Yury Zagorulko	Institute of Informatics Systems, SB RAS, Russia
Victor Zakharov	Federal Research Center "Computer Science and Control" of RAS, Russia
Sergey Znamensky	Institute of Program Systems, RAS, Russia
Mikhail Zymbler	South Ural State University, Russia

DACOMSIN Workshop Program Committee

Igor Abrikosov	Linköping University, Sweden
Toshihiro Ashino	Toyo University, Japan
Keith Butler	Science and Technology Facilities Council, UK
Victor Dudarev	HSE University, Russia
Jennifer Handsel	Science and Technology Facilities Council, UK
Martin Horsch	STFC Daresbury Laboratory, UK
Natalie Johnson	The Cambridge Crystallographic Data Centre, UK
Francesco Mercuri	Consiglio Nazionale delle Ricerche, Italy
Igor Morozov	Joint Institute for High Temperatures of RAS, Russia
Artem Oganov	Skoltech, Russia

Sergey Stupnikov	Federal Research Center "Computer Science and Control" of RAS, Russia
Irina Uspenskaya	Lomonosov Moscow State University, Russia
Yibin Xu	National Institute for Materials Science, Japan

Supporters

National University of Science and Technology MISIS, Moscow, Russia
Thermo-Calc Software AB
Federal Research Center "Computer Science and Control" of the Russian Academy of Sciences, Moscow, Russia
Moscow ACM SIGMOD Chapter

Contents

Data Analysis in Material and Earth Sciences

Information Extraction from Text

Problem Solving Infrastructures, Experiment Organization, and Machine Learning Applications

MLDev: Data Science Experiment Automation and Reproducibility Software

Anton Khritankov$^{(\boxtimes)}$, Nikita Pershin, Nikita Ukhov, and Artem Ukhov

Moscow Institute of Physics and Technology,
Dolgoprudny, Moscow Region, Russian Federation
`anton.khritankov@phystech.edu`

Abstract. In this paper, we explore the challenges of automating experiments in data science. We propose an extensible experiment model as a foundation for integration of different open source tools for running research experiments. We implement our approach in a prototype open source MLDev software package and evaluate it in a series of experiments yielding promising results. Comparison with other state-of-the-art tools signifies novelty of our approach.

Keywords: Experiment automation · Data science · Reproducibility

1 Introduction

The ability to reproduce results and use them in future work is one of the key expectations from the modern data science. Herewith, amount of experimental and empirical papers significantly exceeds the amount of theoretical publications as shown in the analysis [10,18]. Despite the demand and recent progress there are still unsolved problems that hinder further development, reduce trust level and the quality of results [11].

Gundersen et al. [10] study 385 papers containing empirical results from AAAI and IJCAI conferences. More than two third of publications included experiment design, more than a half of the papers included the pseudo code of the algorithm, the problem statement and the training data. At the same time, research questions, purposes of the study, hypotheses tested, source code and detailed analysis of results are often not included in the published papers. Authors indicate that lack of this information significantly influence reproducibility of the research.

Results of the roundtable [22] highlight the reproducibility problem and indicate that a solution of problem requires use of software tools as well as inclusion of topics on experiment design in research training programs in data science.

In 2018 and 2019, organisers of NeurIPS and ICLR conferences [17,18] explored instruments that can be used to increase reproducibility of research. They offered authors the checklists for self-assessment before submitting articles, suggested to provide source code and instructions to reproduce, invited submissions with reproduction of previous researches [18].

© Springer Nature Switzerland AG 2022
A. Pozanenko et al. (Eds.): DAMDID/RCDL 2021, CCIS 1620, pp. 3–18, 2022.
https://doi.org/10.1007/978-3-031-12285-9_1

In this paper we describe our approach to improving reproducibility. We suggest to extract the definition of an experiment from the program code and the paper text and define it in both machine and human-readable form. Such specifications of the experiment should be sufficient to reproduce and automate routine tasks. In order to check our idea, we implement a prototype of MLDev system and test it on several examples.

The main contributions of the paper are as follows. First, we derive quality attributes for data science experiment automation and reproducibility software. Then we propose a new approach to automated execution of experiments based on experiment specification and evaluate it against the requirements and other tools. We also open source the implementation of the propotype MLDev system so that other researches could evaluate our approach.

In the next section, we specify the reproducibility problem. In Sect. 3 we will describe the proposed approach based on the separation of the experiment specification in a standalone artifact. In Sect. 4 we present the results of the empirical evaluation of suggested approach and the analysis of the obtained results. Section 5 includes description of software similar in scope and points out difference with the proposes.

2   Problem Statement

An *experiment* is a procedure carried out to support or refute research hypotheses. Examples of such research hypotheses in data science could be existence of tendencies in data, the choice of model parameters or that one model is not the same as another.

Experiment design is not as simple as it may look. Even a basic data science experiment with random permutation and data splitting into train and test with many such trials exhibits randomization as a design principle to reduce confounding. A structure of the experiment commonly includes the goal, design choices, a list of hypotheses and their acceptance and rejection criteria, source data and expected results. If an experimental procedure is given as a sequence of stages, it is said that an *experiment pipeline* is defined. After running the pipeline, measurements are analysed and conclusions whether hypotheses can be refuted are drawn.

Next, we need to define what is the *reproducibility of experiments*. We use definitions of different types of the reproducibility suggested by NISO and ACM [7]:

- *Repeatability* (same team, same experimental setup). The same measurements can be obtained by the same researcher within the specified error margin using the same procedure and measurement system, under the same conditions. For computational experiments, this means that a researcher can reliably repeat her own computation.
- *Reproducibility* (different team, same experimental setup) An independent group can obtain the same result using the author's own artifacts.
- *Replicability* (different team, different experimental setup) An independent group can obtain the same results using artifacts which they develop completely independently.

Our goal with the MLDev project is to develop software and supporting methodology to help ensure reproducibility, that is, reusability of experiments and their results among researchers. Based on the results of the preliminary literature review, we indicate the identified sources of non-reproducibility [10, 17,18,23]:

- View of the source code as an auxiliary result. Industrial software development methods are not used. These factors result in the code defects and distortion of results as stated by Storer et al. [19].
- Insufficient configuration management, inability to reproduce conditions and procedures of the experiment. This includes execution environment, external dependencies, unavailability of data or source code [23].
- Lack of documentation and insufficient description of experiment [10,20].

Many of these reasons are related to the area of software engineering, a discipline which is not a major for data scientists. Others are related to the willingness to publish the results faster and are most likely caused by the violation of empirical research methodology.

We additionally interviewed heads of data analysis laboratories, academics, students and software developers at Moscow Institute of Physics and Technology, Higher School of Economics and Innopolis university in order to elicit requirements for this type of software. As a result, we highlight the following quality characteristics:

- *REQ1. Extensibility.* Sufficient functionality and extensibility to define and execute a wide range of experiments in a reproducible manner.
- *REQ2. Clarity.* A system should be easy to use and doesn't result in unclear errors or significant increase in time for research.
- *REQ3. Compatibility.* Compatibility with existing libraries and tools for conducting computational experiments.
- *REQ4. No lock-in.* Freedom from risk of impossibility of publishing the obtained results or difficulties during the process.

The complete list of requirements and quality attributes in accordance with ISO/IEC 25010 standard is provided in the Appendix A.

In the next section we propose several architectural decisions, which help overcome the stated reasons of non-reproducibility.

3 Proposed Solution

3.1 Experiment Specification

Following the Model-Driven Development and Language-Oriented Programming approaches [9,21] we address functional extensibility (REQ1) by introducing a separate model for the experiment. The specification of the experiment captures the structure of the experiment and serves as a basis for integration of external tools and data with the user code.

Proposal 1. *Introduce a specification of an experiment as a separate artifact alongside the source code and publication. Provide a core conceptual model of the experiment and means for users to extend it in their experiment specifications.*

The resulting experiment specification allows for the following:

- Gathering all information that is crucial for reproducibility in one place according to a common meta-model provided by the software.
- Use of automated tools for the analysis of the experiment design analysis and generation of results, which is common for language-oriented programming.
- Integration of external tools, both open source and those providing open interfaces.

Let us provide more details on the feature of the proposition.

Instance-Based Composition Model. We propose to define experiment specification as an object-oriented model with instance-based composition instead of class inheritance. In such a model users can add objects with user-defined types and compose new objects from existing ones. The model also supports untyped objects, that is objects use of which does not require specification of a type.

This approach allows users to reuse object state and behaviour by composition, while on the other hand it allows to avoid complexity with polymorphism and virtual inheritance. Indeed, we specifically restrict the use of class-based inheritance to tool and plugin providers who are professional software engineers. Indeed, object-oriented modeling and design competencies are not widespread in the research community and it is still not known whether user-defined class based models are well-suited for experiment design.

Pipeline as a Polyforest. The experiment specification also defines the order of the computation for many hypotheses included in the experiment. Unlike directed computational graphs (pipelines) often used for this purpose, the computational oriented forest (polyforest) seems to fit better. Within an experiment it is necessary to check several hypotheses, algorithms to check each hypothesis are represented as graphs with overlapping vertices. Moreover, an experiment execution context includes the execution order and the usage of services that accompany the execution of the experiment.

The resulting conceptual model is shown at Fig. 1 using the concept map notation. Let's describe each concept described on this map.

Data. In order to ensure better control over the experiment results we include data versioning control, therefore all the inputs and outputs are versioned on every execution.

Stages. An experiment is divided into stages. Inside stages a researcher defines the inputs and outputs, what are the execution parameters. Further, the stages are grouped into pipelines, which makes it possible to effectively use the computational forest model.

Algorithm. The procedure of the experiment is designed so that it tests the hypotheses. It is determined by the source code and the order of stages f execution. The procedure may invoke other algorithms and user source code.

Hypotheses. Hypotheses are tested according to a specific procedure and the results are saved together with the execution logs and reports. Computational forest model enables setting up several experiment scenarios to test more then one hypothesis.

Reports and execution log. Reports and execution log are needed for the control of the runs and analyzing the results of the execution.

Dependencies. Dependnecies include external software and data that needs to be set up before the experiment is run. MLDev creates a virtual environment and installs all the dependencies needed according to specified versions. This process allows to have similar execution environment on another machine and reproduce the experiment.

Services. Services provide more abilities to control experiment execution as well as add user-defined programs that run alongside the experiment with specific functions. For instance, MLDev provides a notification service for a messenger or experiment tracking and logging service. We discuss integrations in more detail below.

3.2 Extensibility Mechanisms

In order to support extensibility (REQ1) provided by the language-oriented programming paradigm and Proposal 1 we apply the microkernel architectural pattern twice. First, we separate MLDev kernel—the base module that is able to interpret and run experiments specifications, and plugins or extras that extend the MLDev functionality. Second, we provide a set of templates that provide artifacts specific to the kind of experiments the user is going to develop and run.

Proposal 2. *Introduce MLDev separate core and plugins that supply implementation to types and objects defined in the experiment specification. Use templates for experiment repositories to provide specific artifacts.*

Microkernel Architecture. Microkernel architecture of MLDev also helps in reducing complexity of the system and run as unnecessary dependencies are not included. Thus we also address REQ2—Clarity. See Table 1 for a list of plugins and open source technologies we use to implement extensions to MLDev base. The implementation of MLDev system as open-source project with dedicated core and template examples help users to adapt system to their needs.

Templates Library. A library of the predefined templates lowers the entry barrier for executing a reproducible experiment and saves the costs spent on preparation and presentation of the results. It does so with the help of partial automation and standardisation. Wherein, template can be adapted for the researchers needs.

The initial template includes the experiment design, the artifacts and reports mockups, which further will be filled with the results of runs. See Fig. 2 for more details. Template can also provide user-defined data types for experiment specification thus extending the original object model. Usage examples will be described further.

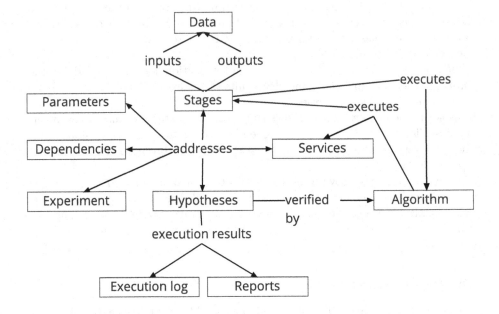

Fig. 1. Conceptual model of the experiment and basic types provided by MLDev. More concepts can be added by plugins and templates. Key: Concept map.

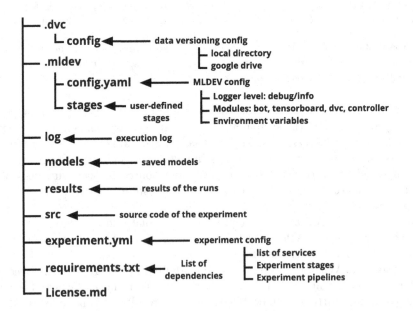

Fig. 2. An example of a template with description of structural elements used by MLDev to provide artifacts for reproducible experiments.

3.3   Open Source Development

Nowadays, a common approach to implement publicly available tools is running an open source project. The open source development approach, if done right, results in a larger participation and cooperation with the target audience. In addition, by running the MLDev project as open source and integrating with other open source tools we address REQ2 by supporting open reviews and continuous community testing, REQ3 through increased number of applications and compatibility tests and REQ4 with an option to fork the project and continue without vendor lock-in. As most open source data science libraries are written in Python, we also implement MLDev in Python thus increasing interoperability (REQ3).

Proposal 3. *Open sourcing the project allows for integration of the other open source technologies that increase automation and reproducibility levels, increase overall quality and functionality.*

Open Source Integration. We leverage available open source tools to extend functionality of MLDev. This is an architectural decision to prefer integration through CLI and API over reimplementing external tools or reusing at the source code level. This provides for lower development costs, better agility and upgradeability although at the cost of efficiency and flexibility. See Table 1 for a list of open source tools used by MLDev in addition to the standard libraries.

Table 1. MLDev uses open source tools and libraries to extend its functionality via plugins.

Feature	Implemented with	Plugin
Configuration management and version control		
Source code	Git	mldev-dvc
Data and results	DVC [1]	mldev-dvc
External libraries	Python venv	mldev
Experiment logging and debugging		
Metrics and measurements	Tensorboard	mldev-tensorboard
Text logs	None	mldev
Notifications	Telegram bot	mldev-bot
Results demo	Flask	mldev-controller
Reproducibility testing		
Unit testing	Pytest, hypothesis	mldev-test*
Code quality	flake8, black	mldev-test*
Reproducibility testing	gitlab-runner CI	mldev
Visualization and presentation		
Notebook support	ipython, Jupyter	mldev-ipython*

4   Experimental Evaluation

4.1   Experiment Design

In order to test our proposals in solving experiment automation and reproducibility problem, we implemented a prototype of the MLDev system [2]. The goal of the experiments is to test whether our design decisions and MLDev implementation are suitable for real-world experiments. That is, we check the feasibility of design so that it could be implemented, and check applicability so that resulting software is suitable for the original experiment automation needs.

In the experiments we check a hypothesis that the MLDev system can be used to develop the experiments from scratch as well as to reproduce experiments conducted earlier. We measure the following metrics and outcomes:

- Was the experiment automation successful?
- Time needed to develop experiment specification for MLDev.
- How similar are results obtained with MLDev and without it?
- Are the results reproducible?

For the experiment we selected one paper that uses MLDev to adapt previously implemented experiment for MLDev with the purpose of reproducibility and another paper by different authors to check whether a third-party experiment can also be adapted to run with the same results via MLDev.

Experiments were conducted during Jan-May 2021 using MLDev versions 0.2.dev3 and 0.3.dev1 respectively on Ubuntu 18 (x64) on a laptop and PC and Google Colab.

4.2   Application to a New Experiment

The article [14] examines the problem of analyzing the quality of continuous machine learning systems, that is systems that use machine learning to implement its major features and that get their predictive models updated over time. The paper presents a simulation experiment for the housing prices prediction problem, in which a feedback loop effect occurs when previously made predictions fall into new training data. The original experiment source code was available as a Jupyter Notebook.

In the process of reproducing the results, the following tasks were solved using MLDev v0.2.dev3:

- Prepared a public repository for an experiment based on the default experiment template.
- The experiment code is extracted from the notebook and is placed according to the structure of the template.
- Implemented a driver program for running an experiment from the command line.
- A description of the MLDev experiment was prepared and the external dependencies of the experiment code were determined for porting to other runtime environments.

– Implemented a script for executing an experiment with tunable hyperparameters to analyze the error margin of the obtained results.

While porting the original source code to the MLDev system, the following issues were identified and resolved in the original experiment:

– An incomplete initialization of the random number generator (affects repeatability).
– An defect in the data update algorithm (does not affect the conclusions of the publication).

The total effort for creating the repository and transferring experiment to the MLDev system were 4.5 h, and another 4 h were required later to refine the experiment, eliminate defects and finalize the visualization of the results. The updated experiment is included in an Arxiv paper [13].

4.3   Reproducibility Study for a Published Paper

Deepak et al. [16] study the problem of predicting relations in knowledge graphs. Authors develop an encoder-decoder architecture that uses a graph neural network as encoder and a convolutional neural network as decoder. The source code is written in Python and run with three console commands with parameters.

When transferring the experiment to MLDev, the following problems were identified:

– An error was detected when re-creating checkpoints used to save models.
– The instructions to run the experiment was too complex and easy to make mistakes, especially when typing them manually.
– There were no parameter configurations for all of the experiment scenarios.

In order to repeat the results and transfer the experiment to MLDev v0.3.dev1, the following steps were taken:

– Eight experiment pipelines have been implemented to form a polyforest. Four of them replicating the original experiment configurations and other four configurations that allow to conduct the experiment with Google Colab [4].
– Resolved a problem with creating checkpoints for saving models.
– A notebook has been created to repeat the experiment.
– A public repository has been prepared with a description of the problem, the result of the experiment and detailed instructions for reproducing the experiment.

The total development effort is five hours and about forty hours were spent on training and testing models. The result of the article was reproduced on the main test dataset FB15k [5]. As a result, MLDev solved the problem of the lack of experimental configurations and made it possible to conveniently reproduce each of the pipelines.

4.4 Analysis and Results

In both cases the experiments were automated and produced the same results which were published by their authors. Also in both cases results can be easily reproducible using single MLDev command.

We were able to confirm that MLDev software is able to run data science experiments with the reproducible results in both cases.

In addition, transferring of the experiments to MLDev helped reveal defects in the source code that might affect the scientific result.

We also confirmed that the experiment specification is extensible enough to accommodate different kinds of experiments: simulation experiments with feedback loops and deep learning experiments.

5 Related Work and Comparison

In this section, we compare MLDev with four related approaches. They are selected in accordance with their popularity in research community and whether they support reproducibility. The manual approach is the most basic that researchers use while preparing the experiment for publication, Jupyter Notebooks are one of the most popular instruments used for ad-hoc research and demonstration of data science experiments [23]. MLFlow and Nextflow are chosen because these projects have existed for a long time, used by researchers in their publications, and these tools are the most representative alternative we could find. Interested reader is also referred to a paper by Isdahl et al. [12] for a review of different experiment automation tools based on the authors reproducibility evaluation framework.

5.1 Command Line Tools

First, we consider the currently most widespread approach to preparing an experiment for publication, in which all documents and source code necessary for reproducibility of the results are prepared manually. Ensuring reproducibility of an experiment is associated with many factors and problems described earlier in Sect. A. These problems include control of the execution environment, determining the order of the experiment execution, etc. The main problem in this case is that even if the researcher manages to provide all the necessary artifacts and describe the procedure for reproducing the experiment, all these components will not be interconnected, as a result the execution of the experiment will not be easy for the researcher who wants to reproduce it. She will need to execute several steps to prepare the environment and install the libraries, then run the scripts for the experiment, then the scripts to evaluate the result. A good example of this approach to experiment design is a source code repository that accompanies the paper by Bunel et al. [6], which we were not able to reproduce without modifying the source code.

In general, this approach is susceptible to most of the problems reported by Pineau et al. [17], such as lack of data for training, incorrect specification of the training process, errors in the code, and so on. If we scale up an experiment to include multiple executions to test multiple hypotheses, the researchers effort increases superlinearly due to interdependencies between pipelines. Similarly, the confidence in correctness and reproducibility of pipelines diminishes.

5.2 Jupyter Notebooks

One of the goals of creating Jupyter notebooks [15] was to ensure reproducibility, but the study by Jiawei W. et al. [23] showed that the presentation of an experiment as a notebook and cells with code brings additional problems related to reproducibility. Nearly 40% of notebooks rely on functions that employ a random number generator, and their execution showed different results from those published by the authors. Less frequent, but also important problems are errors associated with:

- Using time and date functions.
- Using data visualization libraries.
- Lack of input data.
- Lack of dependency control and configuration management.

The main problem associated with the usage of notebooks is the lack of any specified cell execution order. Indeed, notebooks have an option to execute all the cells, but even if the notebook is executed without errors, this will not guarantee the correctness of the results. Thus, designing the experiment as a Jupiter notebook without a clear template and constantly checking the results of the experiment, it is impossible to provide a sufficient level of reproducibility, which is confirmed in a study by Jiawei et al. [23], where they showed that less than 5% of notebooks published in open repositories on GitHub provided the expected results.

Next, we will look at the tools that can be used to build pipelines and aim to provide reproducibility.

5.3 NextFlow

A tool that was developed to run bioinformatics pipelines. Nextflow developers identified the key issues that hinder reproducibility as numerical instability and changes in the runtime when the experiment is ported [8]. NextFlow uses a domain specific language (DSL) and an elaborate meta-model behind the language to define pipelines. A researcher can specify several entry points for the pipeline, thus different scenarios of the experiment can be configured. The problem of numerical instability and changes in the runtime environment is solved by containerizing the stages of the experiment. Therefore, NextFlow can be viewed as a tool to build automated pipelines, while different entry points cannot be considered as specified and easy to use concept for testing several hypotheses.

5.4   MLflow

MLflow is a tool designed for commercial data analysis but can also used by researchers. MLflow developers had three main goals when they were developing their tool [24]:

- Experiment execution tracking to provide for provenance.
- Configuration definition and reuse.
- Model packaging and deployment.

MLflow manages experiments through the concept of a machine learning project. The pipeline is presented as a sequence of stages that are run in a prescribed order. Therefore, it also makes it impossible to test multiple hypotheses within a single experiment configuration. MLflow has many features, but they are more focused on commercial use in production development. This tool can track the execution of the experiment, but due to the fact that it was not developed for research purposes, the inability to record several configurations of an experiment into a single project can become critical when choosing a tool for experiment automation.

5.5   Analysis and Comparison

We considered four common approaches to implement research experiments ranging from an approach without the use of automation tools to the most elaborate solutions that allow you to control the execution of experiments. In accordance with the requirements described in the section where we define the reproducibility problem, we can conclude that tools such as MLflow and NextFlow provide functionality for solving many problems, which researchers face when preparing their experiments for publication. The first two approaches do not provide appropriate automation means out-of-the-box. Regarding the ease of use, the first two approaches may cause a lot of errors and complicate the control of the experiment [17,23]. On the other hand detailed data on the use of MLflow and Nextflow is lacking.

MLflow and NextFlow are simple enough to use for a researcher to conduct an experiment by reading the documentation. Integration with already used tools is partially implemented in MLflow and NextFlow, the ability to write your own additional modules was found only in MLflow and requires special knowledge of building python packages. As for the experiment configuration management, the greatest functionality is provided by MLflow tracking and the ability to specify various execution scenarios in the configuration file. The worst approach to configuration management is, of course, the one without using experiment automation systems at all.

When compared to prototype MLDev implementation, the latter tools are more mature and already received attention from the community, Still, their focus on running data analysis and model deployment (MLflow) or running elaborate bioinformatics pipelines is substantially different from the goals of the MLDev project. Incorporating similar ideas and providing an alternative extensible experiment object model MLDev could be more suitable for a wide range of data science research cases. We summarize the comparison results in Table 2.

Table 2. Comparison of MLDev with other approaches with regards to tool requirements, Sect. A

Approach	Experiment model	Extensions	Project type [3]
Manual	No	CLI/POSIX tools	–
Jupyter notebook	No	Plugins	Monarchist
MLFlow	Yes (fixed)	Plugins, templates	Corporate
Nextflow	Yes (fixed, DSL)	Plugins	Corporate
MLDev	Yes (extensible)	Plugins, model, templates	Community

Thus, we can conclude that at the moment there are tools that partially solve the problem of reproducibility, but there are also aspects in the process of ensuring reproducibility that are not taken into account by current tools, for example, data versioning, or can be significantly improved: experiment configuration for several execution scenarios, custom modules integration.

6 Conclusions

In this paper we study how experiment automation and reproducibility needs of the data science research community could be addressed. Based on literature survey and in-depth interview of the data science professional we propose a novel approach to modeling data science experiments and achieving reproducibility of research.

In general, the creation of the MLDev instrument was motivated by short-comings of the tools available to researchers and the extensive list of unsolved problems that researchers face when it comes to ensuring reproducibility of their experiments [11,17,18].

We implement a prototype MLDev system and apply it to two experiments thus demonstrating feasibility of design decisions and applicability of approach. Comparison with other tools highlights differences in goals of the projects and their target audiences while distinguishing the MLDev approach to extensible experiment model.

A Quality Requirements for Experiment Automation Software

This is a preliminary list of quality requirements for experiment automation and reproducibility software. The requirements are based on series of in-depth interviews of data science researchers, heads of data science laboratories, academics, students and software developers in MIPT, Innopolis university and HSE.

Quality categories are given in accordance with ISO/IEC 25010 quality model standard.

Functionality

- Ability to describe pipelines and configuration of ML experiments.
- Run and reproduce experiments on demand and as part of a larger pipeline.
- Prepare reports on the experiments including figures and papers.

Usability

- Low entry barrier for data scientists who are Linux users.
- Ability to learn gradually, easy to run first experiment
- Technical and programming skill needed to use experiment automation tools should be lower than running experiments without it.
- Users should be able to quickly determine the source of the errors.

Portability and Compatibility

- Support common ML platforms (incl. Cloud Google Colab), OSes (Ubuntu 16, 18, 20, MacOS) and ML libraries (sklearn, pandas, pytorch, tensorflow...)
- Support experiments in Python, Matlab
- Run third-party ML tools with command-line interface

Maintainability

- Open project, that is everyone should be able to participate and contribute.
- Contributing to the project should not require understanding all the internal workings.
- Should provide backward compatibility for experiment definitions.

Security/Reliability

- Confidentiality of experiment data unless requested by user otherwise (e.g. publish results).
- Keep experiment data secure/safe for a long time

Efficiency

- Overhead is negligible for small and large experiment compared with the user code.

Satisfaction and Ease of Use.

- Must be at least as rewarding/satisfactory/easy-to-use as Jupyter Notebook.
- Interface should be similar to other tools familiar to data scientists.

Freedom from Risk

- Using experiment automation software should not risk having their projects completed and results published.

References

1. Data version control tool (dvc). https://dvc.org. Accessed 14 June 2021
2. MLDev. An open source data science experimentation and reproducibility software. https://mlrep.gitlab.io/mldev. Accessed 14 June 2021
3. Berkus, J.: The 5 types of open source projects. https://wackowiki.org/doc/Org/Articles/5TypesOpenSourceProjects. Accessed 14 June 2021
4. Bisong, E.: Google colaboratory. In: Building Machine Learning and Deep Learning Models on Google Cloud Platform, pp. 59–64. Springer. Apress, Berkeley, CA (2019). https://doi.org/10.1007/978-1-4842-4470-8_7
5. Bordes, A., Usunier, N., Garcia-Duran, A., Weston, J., Yakhnenko, O.: Translating embeddings for modeling multi-relational data. In: Neural Information Processing Systems (NIPS), pp. 1–9 (2013)
6. Bunel, R., Hausknecht, M., Devlin, J., Singh, R., Kohli, P.: Leveraging grammar and reinforcement learning for neural program synthesis. In: International Conference on Learning Representations (2018). https://openreview.net/forum?id=H1Xw62kRZ
7. Chue Hong, N.: Reproducibility badging and definitions: a recommended practice of the national information standards organization. Nat. Inf. Stan. Organ. (NISO) (2021). https://doi.org/10.3789/niso-rp-31-2021
8. Di Tommaso, P., Chatzou, M., Floden, E.W., Barja, P.P., Palumbo, E., Notredame, C.: Nextflow enables reproducible computational workflows. Nat. Biotechnol. **35**(4), 316–319 (2017)
9. Dmitriev, S.: Language oriented programming: the next programming paradigm. JetBrains onboard **1**(2), 1–13 (2004)
10. Gundersen, O.E., Gil, Y., Aha, D.W.: On reproducible AI: towards reproducible research, open science, and digital scholarship in AI publications. AI Mag. **39**(3), 56–68 (2018)
11. Hutson, M.: Artificial intelligence faces reproducibility crisis (2018)
12. Isdahl, R., Gundersen, O.E.: Out-of-the-box reproducibility: a survey of machine learning platforms. In: 2019 15th International Conference on eScience (eScience), pp. 86–95. IEEE (2019)
13. Khritankov, A.: Analysis of hidden feedback loops in continuous machine learning systems. arXiv preprint arXiv:2101.05673 (2021)
14. Khritankov, A.: Hidden feedback loops in machine learning systems: a simulation model and preliminary results. In: Winkler, D., Biffl, S., Mendez, D., Wimmer, M., Bergsmann, J. (eds.) SWQD 2021. LNBIP, vol. 404, pp. 54–65. Springer, Cham (2021). https://doi.org/10.1007/978-3-030-65854-0_5
15. Kluyver, T., et al.: Jupyter Notebooks-a Publishing Format for Reproducible Computational Workflows, vol. 2016 (2016)
16. Nathani, D., Chauhan, J., Sharma, C., Kaul, M.: Learning attention-based embeddings for relation prediction in knowledge graphs. In: Proceedings of the 57th Annual Meeting of the Association for Computational Linguistics, pp. 4710–4723. Association for Computational Linguistics, Florence, Italy, July 2019. https://doi.org/10.18653/v1/P19-1466, https://www.aclweb.org/anthology/P19-1466
17. Pineau, J., Sinha, K., Fried, G., Ke, R.N., Larochelle, H.: ICLR reproducibility challenge 2019. ReScience C **5**(2), 5 (2019)
18. Pineau, J., et al.: Improving reproducibility in machine learning research (a report from the Neurips 2019 reproducibility program). arXiv preprint arXiv:2003.12206 (2020)

19. Storer, T.: Bridging the chasm: a survey of software engineering practice in scientific programming. ACM Comput. Surv. (CSUR) **50**(4), 1–32 (2017)
20. Trisovic, A., Lau, M.K., Pasquier, T., Crosas, M.: A large-scale study on research code quality and execution. arXiv preprint arXiv:2103.12793 (2021)
21. Voelter, M.: Fusing modeling and programming into language-oriented programming. In: Margaria, T., Steffen, B. (eds.) ISoLA 2018. LNCS, vol. 11244, pp. 309–339. Springer, Cham (2018). https://doi.org/10.1007/978-3-030-03418-4_19
22. Vorontsov, K., Iglovikov, V., Strijov, V., Ustuzhanin, A., Khritankov, A.: Roundtable: challenges in repeatable experiments and reproducible research in data science. Proc. MIPT (Trudy MFTI) **13**(2), 100–108 (2021). https://mipt.ru/science/trudy/
23. Wang, J., Tzu-Yang, K., Li, L., Zeller, A.: Assessing and restoring reproducibility of Jupyter notebooks, pp. 138–149 (2020)
24. Zaharia, M., et al.: Accelerating the machine learning lifecycle with MLflow. IEEE Data Eng. Bull. **41**(4), 39–45 (2018)

Response to Cybersecurity Threats of Informational Infrastructure Based on Conceptual Models

Nikolay Kalinin[1]([✉]) [iD] and Nikolay Skvortsov[2] [iD]

[1] Faculty of Computational Mathematics and Cybernetics, Moscow State University, Moscow 119991, Russia
Kalinin-na@yandex.ru
[2] Institute of Informatics Problems, Federal Research Center "Computer Science and Control" of the Russian Academy of Sciences, Moscow 119333, Russia

Abstract. Response to the threats of information security in conditions of modern organization with a large infrastructure is an area with emergency loaded intensity of the data usage. For a successful exposure and the prevention of computer attacks the construction of complex models of the events and infrastructure is required. In this work, the question of the applicability of ontological models and reasoning for supporting response process is examined. On the basis of built ontology, practical use cases are demonstrated.

Keywords: Ontology · Reasoning · Cybersecurity · Threat response

1 Introduction

In recent years, the continuous and rapid development of computer systems led the growing information security risks. Due to the tangled architecture of modern systems, traditional methods of protection based on the security perimeter have lost sufficient part of their effectiveness. The most promising trend in enterprise information security for big organizations is Security Operations Centers with broad competencies, many security and monitoring tools. Such centers can aggregate the whole related information in one place and realize really challenging conception like zero trust model [18]. But the implementation of a centralized security monitoring model is associated with a high workload, and the most critical process, in the case, is the cyberthreat response. In this paper we show ontology-based approach to formalize and automation of cyberthreat response process. As noted in the work [22] ontology is already one of the fixed assets of realization of the large systems of information security. The main advantages of formal models: strict interopretability and high-level semantic integration of disparate data sources. In cybersecurity area it also allows to improve inter-team interaction and to automate the process of maintaining internal databases by integrating with open dictionaries and taxonomies. Tools of exposure threat on

A. Pozanenko et al. (Eds.): DAMDID/RCDL 2021, CCIS 1620, pp. 19–35, 2022.
https://doi.org/10.1007/978-3-031-12285-9_2

the basis of formal models can allow not only to identify and classify threats but also to effectively produce reliable and interpreted decisions for their removal. A brief overview of related work is given in Sect. 2. Section 3 describes the structure of the model, the selected use cases, implementation issues and a brief overview of the popular open standards and taxonomies applied within the model. Section 4 demonstrates the capabilities of the model via use cases.

2   Related Works

A more detailed analysis of related works was given in our previous article [12], so here we will limit ourselves to a review of some new publications and a general conclusion about the state of the subject area. A number of authors continue to focus on the redevelopment of existing standards, especially the hierarchically represented CWE and CAPEC [4], and tried to construct a universal intermediate ontology [6]. But the most interesting, in our opinion, is the applied approach presented in [30] for graphical modeling of cybersecurity assessment, as well as for [27] aggregation of security logs. In [27], a complex view was also made for the compatibility of security logs. In the field of vulnerability management, ontologies demonstrate rapid development, for example in [29] modeling of corporate threats for the shown and in [20] reasoning is successfully used to analyze and classify vulnerabilities. The importance of infrastructure modeling is shown in [2], where not only computational, but also physical infrastructure was analyzed. An advanced approach to the choice of countermeasures is used in [25]. Finally, a complex overview of existing ontologies was presented in [16]. Before traditional gap analysis, it is important to say that the rapid development of this field should very soon lead to serious practical results, especially for terminologies that provide a common semantic basis. The most common problem for practical use of security ontologies is difficult implementation and mistrust of practitioners. In our opinion, in addition to the lack of integration for the operational ontologies highlighted in [16], there are a number of problems. First, the internal infrastructure is poorly represented in security ontologies. Second, most of them build more for managing people, rather than managing information systems. Finally, the reasoning capabilities were shown only for small area analysis, both for SQL injections in [5] and for WAF rules in [1]. Thus, we hope that our view of the threat response process, driven by the reasoning abilities at every step for real-world tasks, will partially close this shortcoming.

3   Model Overview

We used two main initial data types. The first represents high level security concepts and presented in open sources, a detail analysis of which is given below. The second is low level data, collected from enterprise security and inventory systems. First version of our model, presented in [12], was built as an extension for existing ontology UCO [26], but thereafter we decide save minimal integration with external ontologies and concentrate on applied level. Thus, we use reversed middle-out methodology for built our model. First, we define main high-level concepts, based on common term and relationships between them in threat response point of view, most of them matched with cocoa [17] concepts. Then we define classes and relationships for our data description. And finally, the middle layer constructs to describe links between previous two, based on the identified use cases.

3.1 Use Cases

We highlighted four basic use cases for built ontology model: attack detection and classification, risk assessment, incident generation and selection of counter-measures.

Attack Detection and Classification. Incident response analytic, as noted above, ordinary works with varied security logs, generated by number of security systems. Some of such systems include internal attack detection rules, but this rules very often require a revision, moreover, used system do not cover all possible vectors. Therefore, security teams have to build custom rules. Especially rules customization is important for the zero-trust model. For formalization this process we suggest to use the logical criteria how to define attack class based on log messages.

Risk Assessment. When an evil action is detected we need to determine how dangerous it is - estimate the risk. The basement for the assessment is adopted CVSS methodology and the most interesting phase is the environmental metrics calculation, which include determination of the attack surface. After finding all affected nodes we can use the inventory databases for finding their criticality class and finally estimate impact on confidentiality, integrity and availability.

Incident Generation. The most import incident response process - process of finding relevant data. We call it here the incident generation, because of method we use: in our ontology we define the special interface-like object, which is linked with all required data. It simplifies program integration and help us partly to avoid difficult SPARQL queries using.

Selection of Countermeasures. Main goal of incident response is to stop intruder and recover damaged systems thus selection of the right countermeasure is really important. For every attack we can define three stage countermeasure: localization, mitigation and recovery. Selection a right one is based on inferred attack class, related data, risk level and infrastructure knowledge's. Formalization of the countermeasures can also simplify their automatic execution, as an example we built the osquery extension integrated with our ontology.

3.2 Implementation

We developed our model with using Python and Owlready framework, which support ontology-oriented programing: effective creation, querying and manipulation of the ontology content. Owlready has several advantages: It's easy to use, has support for runtime T-BOX changing, helps to avoid difficult SPARQL queries [13]. Our ontology can be divided on four logical parts. The first and the most significant serves for event description. It based on main class "Event". Events, that have signs of suspicious activity called Attacks and defined as sub-classes. Thus, attack detection is based on the event instances classification. Every Attack is linked with Threat. Threat description describe predetermined attack characteristics: stage, suspended platforms, goal and based on ATT&CK matrix [23]. The second contains networks concepts: Node, Interface, Segment, Firewall rule and additional information about endpoints: installed and active software, criticallity class, business role, relations with other business systems. The third serves for countermeasures description. In addition to the three types of countermeasures already mentioned above: Localization, Recovery and Miti-gation this part contains descriptions of specific infrastructure mechanisms for their implementation. In particular, if an interactive connection to the attacked node is required, the descriptions of this countermeasure class contain instruc-tions on the accounts and access points that must be used. The last one is needed for support of the analyst's work and includes interface-like classes for describing current infrastructure state. This classes can be used after response process for post-mortem analytic and security improvements. We also developed program modules for creating Individuals from data. The following data sources are used: audit and osquery[1] framework monitoring system logs, firewall rules, network inventory databases, endpoints inventory databases. The list of sources advises the practical implementation of the model presented in [17]. In addition to the data described above, which are hereinafter referred to as operational, some of the instances also arise as a result of integration with open standards and taxonomies: CVE, CWE and Exploit DB.

It is important to note that the practical applicability of the model depends on the quality of the initial knowledge bases. Despite the fact that the defined restrictions in the event-classes (for example, the condition of uniqueness of the identifier or strict compliance between the ip and device name) allow us to detect errors that inevitably occur in data, these descriptions themselves require careful verification. The issues of implementing such verification are not considered in this paper, but successful examples of implementing ontological systems based on highly formalized sources (PyMedTermino2 [14]) allow us to consider this problem solvable.

3.3 External Standards and Taxonomies

In previous section we noted some of the widespread standards and classifica-tions: CVE, CWE, CAPEC, ATT&CK, here we will dwell on them in more detail.

CVE is the primary standard for filing and uniform naming of vulnerabili-ties. Despite a significant amount of research [10,24], fully automated semantic interpretation of semi-structured CVE descriptions is a serious problem. It is also not possible to manually process the data from the CVE list, due to its size, so we used a hybrid approach: Standard data contained in the records: text

---

[1] https://osquery.io.

description, vulnerable software products and CVSS risk level are obtained automatically using CVE Changelog, and detailed ontological descriptions, including exploit descriptions from linked Exploit DB records, are generated by the analyst only for the most relevant vulnerabilities.

Since not all software weaknesses and especially misconfigurations are described by CVE identifiers, when creating your own definitions, it is also possible to use the classification of CWE software flaws, for which records are described similarly to CVE, but preserving the original hierarchical structure [15].

The most actual taxonomy for us is CAPEC. The object of systematization in it are attack patterns, that is, descriptions of common elements and methods used in attacks on vulnerable components. CAPEC uses a hierarchical approach similar to CWE. The ATT&CK matrix became the development of CAPEC in terms of classifications of actions of intruders. ATT&CK is based on real observations and contains, not only descriptions of common methods of attackers, but also countermeasures. The key practical difference between ATT&CK and CAPEC is the use cases. CAPEC focuses on describing operating techniques in software products while maintaining an essential link to the CVE inventory. ATT&CK provides a comprehensive view of the attacker's behavior in the network infrastructure as a multi-stage attack is implemented. Also, CAPEC is more formal and ATT&CK is more practical oriented. The approach we describe in many respects repeats the logic of CAPEC and ATT&CK [23]. The attack classes of our model correspond to the ATT&CK tactics, and for many definitions the corresponding CAPEC ID is given.

As mentioned above, ATT&CK contains some classification of countermeasures, but only in additional information format. There are no other generally accepted classifications or enumerations for countermeasures, although some developments are presented in a number of works: For example, some classification is given in [8,9] also provides a more detailed classification for countermeasures against network attacks, and [11] shows the importance of maintaining the relationship with countermeasures when building a threat classification. Due to the lack of a generally accepted industrial classification, this work uses a somewhat extended approach, similar to the classification proposed by MITRE ATT&CK: each ATT&CK technique corresponds to countermeasures that can actually be used in the infrastructure (see table in Appendix).

In addition to the listed classifications of global concepts, the ontological approach also allows the use of local classifications for the described resources. An example of such a classification, which was implemented in the developed model, is the classification of Linux system commands, the wide possibilities of using which for detecting attacks are demonstrated in [19].

4   Use Cases of Model Application

4.1   Attack Classification

Attacks are detected and classified by creating sub-classes of the Attack class. In "equivalent to" construction we can use the entire completeness of the descriptive logic, including using other classes, which allow us formal define detection rules, including rules based on black and white lists. For simple example, we define simple rule for exploitation new vulnerability in sudoedit utility as Explotation_CVE_2021_3156, based on vendor notification.

```
Class: <Explotation_CVE_2021_3156>
    EquivalentTo: <SyscallEvent>
        and (<has_argument> some
        xsd:string[minLength "20"^^xsd:nonNegativeInteger])
        and (<has_exe> value "/usr/bin/sudo")
        and (<has_cmd_flag> value \-s")
        and (<has_cmd> value \sudoedit">
    SubClassOf: <Attack>
```

Thus, log string below would be parsed into Instance of Event and then inferred as Attack (Explotation_CVE_2021_3156).

```
25.03.2021 15:25; srv-deb-08;10.117.0.5;1000;
1000;/usr/bin/sudo;sudoedit -s
AAAAAAAA..\  BBBBBBB...\;
4919;790;4261;yes;0;1000;0;1000;0;1000;0;2;
rootcmd;6;1000;unconfined;59;pts1
```

```
Individual: <cevent9>
Types: <SyscallEvent>
    Facts:
    <has_argument> "-s AAAAAAAAAAAAAAAAAAAAAAAAAAAA...\ ",
    <has_command>  "sudoedit",
    <has_exe>  "/usr/bin/sudo",
    <has_payload>  "sudoedit -s AAAAAAAAAAAAAAAAAAAAAAAAAAAA...\",
    <has_timestamp>  "25.03.2021 15:30",
    <has_type>  "syscall"
```

The second example demonstrates how black and white list can be use in definitions. We have well known lists of hacker tools and make a special class (part of command taxonomy).

```
Class: <Hacktool >
    EquivalentTo: <LinuxCommand>
        and (<has_argument> only
        {'metasploit', 'nmap', 'masscan'...})
    SubClassOf: <LinuxCommand>
```

Then attack description can use this class, but class themselves can be managed separately and obtained from external sources.

```
Class: <HacktoolUsage>
```

```
EquivalentTo: <SyscallEvent>
         and (<has_command> some <Hacktool>)
SubClassOf: <Attack>
```

Despite the fact that the detection of attacks through the definition of signature rules is still the most common industrial method, methods based on the detection of anomalies are gaining more and more popularity. Anomaly detection, because of needed of interpretation, very often based on mining rules of normal behavior, for example ARL [3]. Such rules, almost always can be defined as implication

$$X \rightarrow Y \tag{1}$$

where X and Y is subset of transaction attributes. For demonstration we mined normal behavior rules for ssh connections:

```
('src_ip', [10.110.110.1, 10.112.111.2, 10.53.42.65]) =>
('dst_ip', ['10.82.16.13'])
('username 'system_critical_admin') =>
(dst_ip, ['10.18.18.18', '10.18.18.19', '174.18.9.6',])
( 'dst_ip': ['10.182.7.35']) =>('src_ip', '10.3.6.17')
```

Rules can be defined in ontology. Left operand describes domain of the rule and might be any Event subclass, even Event themselves for global rules.

```
Class <NodeRule1Set>:
EquivalentTo: <Node> and <has_ip>  only
{10.110.110.1, 10.112.111.2, 10.53.42.65})
SubClassOf: <Node>

Class: <EventDefinedByRule1>
   EquivalentTo: <Event> and (<has_source> some <NodeRule1Set>)
   SubClassOf: <Event>

Class: <EventDefinedByRule2>
   EquivalentTo: <Event> and (<has_username>  value
   "system_critical_admin")
   SubClassOf: <Event>

Class <NodeRule3Set>:
   EquivalentTo: <Node> and <has_ip> only {10.182.7.35})
SubClassOf: <Node>

Class: <EventDefinedByRule3>
   EquivalentTo: <Event> and (<has_target> some <NodeRule3Set>)
   SubClassOf: <Event>
```

A right operand is a condition for matching the rule:

```
Class <NodeRule1SetR>:
EquivalentTo: <Node> and <has_ip> only {10.82.16.13})
SubClassOf: <Node>

Class <EventBreakingRule1>:
EquivalentTo: <EventDefinedByRule1> and not <EventFitByRule1>
SubClassOf: <EventDefinedByRule1>

Class <EventDefinedByRules>:
EquivalentTo: <EventDefinedByRule1> or <EventDefinedByRule2>
or <EventDefinedByRule3>
SubClassOf: <Event>

Class <EventFitRules>:
EquivalentTo: <EventFitByRule1> or  <EventFitByRule2>
or  <EventFitByRule3>
SubClassOf: <Event>

Class <EventBreakingRule>:
EquivalentTo: <EventDefinedByRule> and not <EventFitByRule>
SubClassOf: <Event>

Class: <EventFitByRule3>
    EquivalentTo: <Event> and (<has_source> some <NodeRule3Set>)
    SubClassOf: <EventDefinedByRule3>
```

And rule violation can be defined either as a negation of the intersection of classes, or as in a more general representation, based on the enumeration of all classes of rules.

This definition can be easily built automatically due to logic likeness. The use of an ontological model for implementing rules allows us to further improve the degree of their interpretability, due to the possibility of additional descriptions and establishing an identity relationship between classes. The rule description does not contain information about the reasons for this particular set of nodes of the left operand. As a result of reasoning a correspondence is established between the class describing the left operand and the system class from the inventory data: EventDefinedByRule = BrokerSystemNodes Fig. 1.

The meaning of the formal classification of events is not only in the actual detection of the attack. The definitions of specific attacks also include references to the threats implemented by this attack, the affected nodes, and other descriptions. The link to threats implies an indirect link to countermeasure, examples of which are given below.

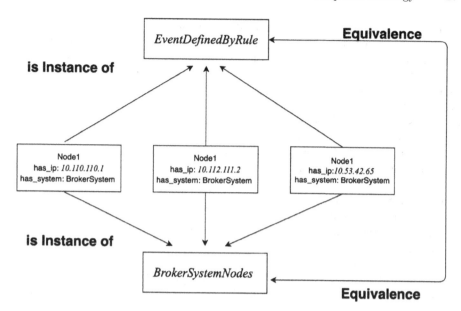

Fig. 1. Class equality

4.2   Network Nodes Reachability and Risk Estimation

To describe the relationship between events and infrastructure elements, descriptions of nodes (Node), their network interfaces (Interface), software installed on them (Software), including software active on network ports (Services) are implemented.

The model also implements classes for describing the logical structure of the network-segments (Segment) and firewall rules (Rule). As shown in [21], network descriptions based on SWRL rules can be used to establish the reachability relationship between two nodes, this paper uses an approach similar to RMO, but does not involve the use of rules. From the point of view of the response team analyst, the most important is the question about the nodes affected by the attack. The event individual received at the attack detection step already contains information about the directly affected nodes (the has_target relation).

```
Class: <AttackedNode>
    SubClassOf: <Node>
    EquivalentTo:  <Node> and (<is_tagret_for> some <Attack>)

Class: <AttackedInterface>
    SubClassOf: <Interface>
    EquivalentTo:  <Interface> and
    (<belongs_to_node> some <AttackedNode>)
```

However, if the recorded event indicates that the node has been compromised, it is also necessary to determine the list of resources to which the attackers obtained network access as a result of the compromise. Let's define the auxiliary classes of the segment that is accessed (Compromised Segment) and the rules that can be used by an attacker. (AttackerAffectedRule)

```
Class: <CompromisedSegment>
    EquivalentTo: <Segment> and
    (<has_connected> some <AttackedInterface>)
    SubClassOf: <Segment>

Class: <AttackerAffectedRule>
    EquivalentTo: <Rule>and
    (<has_source> some <AttackedInterface>)
    SubClassOf: <Rule>
```

Now, naturally, the class of interfaces available to the attacker and the corresponding class of nodes can be defined.

```
Class: <RuleAllowedSurface>
    EquivalentTo: <Interface> and
    (<allowed_by> some <AttackerAffectedRule>)
    SubClassOf: <Interface>

Class: <AttackerAffectedPortRule>
    EquivalentTo: <Port> and
    (<port_allowed_by> some <AttackerAllowRule>)
    SubClassOf: <Port>

Class: <PortRuleSurface>
    EquivalentTo: <Port> and
    (<open_on_interface> some <RuleAllowedSurface>)
    SubClassOf: <Port>

Class: <PortSegmentSurface>
    EquivalentTo: <Port> and
    (<open_on_interface> some <SegmentSurface>)
    SubClassOf: <Port>

Class: <ServiceRuleSurface>
    EquivalentTo: (<Service> and
    (<has_open_port> some <AttackerAffectedPortRule>))
    or (<has_open_port> some <PortRuleSurface>)
    SubClassOf: <Service>

  Class: <ServiceSegmentSurface>
```

```
EquivalentTo: <Service> and
(<has_open_port> some <PortSegmentSurface>)
SubClassOf: <Service>
```

```
Class: <AffectedService>
    EquivalentTo: <ServiceRuleSurface> or <ServiceSegmentSurface>
    SubClassOf: <Service>
```

Firewall rules usually restrict access to certain ports or protocols as well, and the above example can be used with a little adaptation in this case. For attacks involving further lateral movement of the attacker, it is possible to expand the list of nodes related to the affected ones by introducing a transitive relation between the objects of the presented classes.

The list of affected nodes constructed in the described way is used to assess the danger of an event. As a demonstration of the capabilities of the ontological model, two approaches to risk assessment will be described. The first involves an implementation similar to the Vulnerability Scoring Standard (CVSS). The second is based on local prioritization policies. Despite the fact that CVSS is designed for vulnerability assessment, it is easy to see that this approach can easily be generalized to assess vulnerability exploitation events. To do this, the descriptions of the attack classes must contain an indication of the value of the base and time risk, as well as the modified base metrics. To get the environmental risk, in the built ontology the relations has_ availability_security_ requirements, has_integrity_security_ requirements, has_confidentiality_security_requirements are defined, its generate 27 classes of criticality of resources. The metric counting method is implemented as an imperative owlready method. More interesting, from the point of view of the ontological implementation, is the second example, since it is the use of local risk assessment policies that is standard for industrial cybersecurity teams. We will call the nodes that are the direct target of the attack (is_target for_some Attack) the attacked nodes, and the nodes and services that are accessed as a result of the attack, the affected nodes. The event risk level is determined based on the degree of criticality of the attacked and affected nodes, the degree of danger of the attack class, which can be determined by the analyst or derived from the relationship with the ATT&CK tactics (the Threat class of the model). In addition, the existence of known vulnerabilities is considered for the affected nodes and services.

Classes of affected, attacked resources, as well as classes of attacks and systems are defined, as partially shown in the last section. Vulnerable elements can be identified by linking to the CVE. Therefore, no additional logical inference is required to calculate the risk. Pseudo code shown below.

```
# Affected is an example. Attacked and Vulnerable used in same way
AffectedHighResource = (AttackedNode | AttackedService)
& belong_to (System & has_criticality High)
```

```
AffectedMidResource = (AttackedNode | AttackedService)
& belong_to (System & has_criticality High)
AffectedLowResource = (AttackedNode | AttackedService)
& belong_to (System & has_criticality High)

#Get custom criticality score
AffectedHighResource.has_score = 100

# Get number for high affected node and services
Len(AffectdedHighResource.instances())
...
# Get score for vulnerable
For VulnerableResources.instances(): score =
resources_instance.cve.score  * resource_criticality_score
...
Calculate_score(metrics)
```

The model allows us to combine at one level of abstractions knowledge about threats and vulnerabilities, knowledge about the state and structure of the information infrastructure and knowledge about business processes and their criticality. As the result, the analyst, together with the risk assessment, receives comprehensive information about the affected objects. This information becomes the basis for the choice of countermeasures.

4.3   Countermeasures

When attacks are detected, three individuals of countermeasure classes arise: Localization, Recovery, Mitigation. Further refinement of actions is the classification of these individuals. Selected subclasses advise the answers to these questions. Thereby model include three classifications for contractions by action type, by action object, by action tool. On Fig. 2 an example of classification for localization, based on our information is shown. It is easy to see that based on the above scheme, 20 finite localization classes are defined.

Localization measure class for the vulnerability *Explotation_CVE_2021_3156* exploitation event can be defined by this reasoning: the vulnerability belongs to the privilege escalation class - we need to block access, since the attack is local, the target is to access a specific system, the tool for blocking it is the operating system interface, exploitation of the vulnerability cannot be unintentional, it also requires notification of internal authorized persons. All selected parent classes contain verbal description of required actions, but some of them also can include imperative methods to call external systems functions for automatic response.

4.4   Incident Generation

Direct interaction between the user and the formal model is time-consuming, so to implement such interaction, the model implements the Report class and

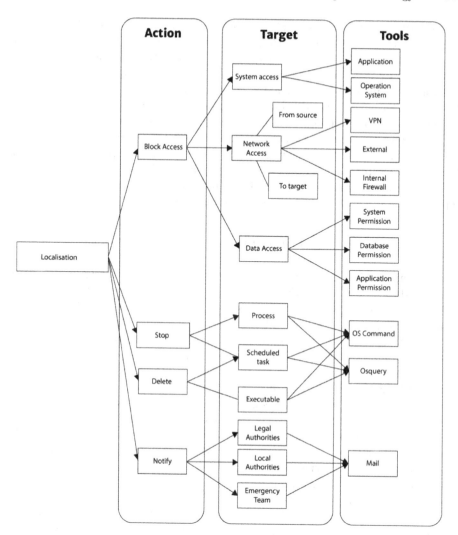

Fig. 2. Countermeasure selection

the Incident Report class. A report is an incident report if there are attacks among the related events. We call this classes interface classes here, since they are equally important in the form of an OWL class and a python class. They are designed to aggregate and present all the data necessary for the response. The composition of these classes may differ significantly, depending on the user needs. As an example, in the constructed model, the main functions of these classes include: uniform presentation of information about events and their properties, procedural calculation of statistical metrics, presentation of information about the general state of the infrastructure, and the formation of an informal report. Because the user may not know anything about the internal structure of the formal model, the incident class descriptions contain properties for basic needs, which are compositions (PropertyChain) of "internal properties" needed to access specific elements. For example, the properties of related events are represented by the chain *has_event * event_property_name*, and the hosts of the fixed attacks are represented as the values of the property *has_target_node = has_event * has_target*. In addition to the interface for accessing knowledge base objects, the report also contains statistics, such as the number of events during the reporting period, the number of nodes affected by attacks, the number of attacks of various types, and the number of vulnerable copies of software in the infrastructure. These aggregates are calculated using python methods [13]. An informal report is created based on the interface classes, any other data representation (web form, tables, etc.) can also be created.

5   Conclusions and Directions for Further Work

The implemented model allows us to formalize a wide range of tasks for responding to cyber threats: detection, risk estimation, countermeasures selections. Infrastructure object was linked with security classes. Reasoning abilities were combined with imperative methods, and "scary" ontological concepts were hidden behind interface classes. Despite its practical orientation, the model construction supports integration with reference ontologies. In our previous paper we described how such model might be integrated with other ontologies. Without detracting from the importance of such integration, it is worth noting that the practical use of external ontologies, just like LOD, is fraught with risks of resource unavailability, since most of them have weak support. The main advantage of such integration is inter-team interoperability. We think, in case of such a need, it better to use "weak" integration throw building association between classes. Full-fledged practical implementation requires serious work on the content and adaptation of the ontology. The described model is only a framework for implementing the corporate knowledge base. It has several places to fill in. First, the conditions for classifying events must be selected and developed. As we show, some of them can be built automatically, but a significant part comes from informal sources and needs to be manually implemented. Second, in some cases, the network model can be changed by adding new network entities, such as, for example, NAT rules. Third, the countermeasure must be adapted with real tools. Finally, the format of the report object is the matter of opinion. Important possibilities are also presented in the area of integration with other security systems. For example, [1] successfully uses the web application firewall ontology model. The same results are shown in [5] for SQL injection detection. Other examples are IDS systems [28], container security modules, data leak prevention systems, and even Internet of things security modules [7]. For the countermeasures, integration with infrastructure management tools such as Ansible or Puppet also possible. In our opinion, the declarative semantics of an Ansible playbooks is of particular interest. Separately, we would like to emphasize that the process of determining the relationships between objects of the real world for their ontological implementation can be very time-consuming. In some cases, this problem can be completely solved by an ontologist, but automated tools are required for the real implementation. Thus, the development of approaches to automated knowledge extraction from databases, concepts and principles of their verification, methodology for evaluating the quality of automated tools are required.

Our future direction is filling in the ontology for further application development, both manual and automated. The ontology-based application integrated with some of popular event management tool might allow expand the view of the ontologist with a million of practical use cases and become a basis for competition with traditional systems.

Appendix A

Tactic	Localization	Recovery	Mitigation
Reconnaissance	Block access	Compromised data update	Privilege limitation, sensitive data management
resource development	Block compromised resources	Delete compromised resources	Resource management, leak hunting
Initial Access	Block access vector (credentials, network, web application)	Delete compromised resources, compromised data update	Update vulnerable component, isolate applications
Execution	Stop/delete execution resources (processes, code, scheduled tasks)	Reverting execution results	Update vulnerable component, Privilege limitation,
Persistence	Stopping persistence items (job, software e.t.c.)	Delete persistence item (job, software e.t.c.)	Privilege limitation, System integrity control
Privilege escalation	Stopping privilege items	Delete privilege items, deleting compromised resources	Privilege limitation, System integrity control
Defense evasion	Block access vector, deleting compromised system resources	Defense recovering	Privilege limitation, system integrity control, defense mechanism control
Credential access	Block compromised access	Update compromised resources	Privilege limitation
Lateral movement	Block compromised access	Update compromised resources, delete malicious resources	Privilege limitation, updating vulnerable component
Collection	Block data access	Update compromised data, Delete compromised resources	Data access limiting
Command and control	Block compromised system	Remove malicious resources	Privilege limitation, system integrity control
Exfiltration	Block compromised system, block data access	Update compromised data, delete exfiltration resources	Data access limiting

References

1. Ahmad, A., Anwar, Z., Hur, A., Ahmad, H.F.: Formal reasoning of web application firewall rules through ontological modeling. In: 2012 15th International Multitopic Conference (INMIC), pp. 230–237. IEEE (2012)
2. Canito, A., Aleid, K., Praça, I., Corchado, J., Marreiros, G.: An ontology to promote interoperability between cyber-physical security systems in critical infrastructures. In: 2020 IEEE 6th International Conference on Computer and Communications (ICCC), pp. 553–560. IEEE (2020)

3. Cheng, M., Xu, K., Gong, X.: Research on audit log association rule mining based on improved Apriori algorithm. In: 2016 IEEE International Conference on Big Data Analysis (ICBDA), pp. 1–7 (2016). https://doi.org/10.1109/ICBDA.2016.7509792

4. Dimitrov, V., Kolev, I.: An ontology of top 25 CWEs (2020)

5. Durai, K.N., Subha, R., Haldorai, A.: A novel method to detect and prevent SQLIA using ontology to cloud web security. Wireless Pers. Commun. 1–20 (2020)

6. Gladun, A., Khala, K., Subach, I.: Ontological approach to big data analytics in cybersecurity domain. Collect. Inf. Technol. Secur. **8**(2), 120–132 (2020)

7. Gonzalez-Gil, P., Martinez, J.A., Skarmeta, A.F.: Lightweight data-security ontology for IoT. Sensors **20**(3), 801 (2020)

8. Gregg, M., Watkins, S., Mays, G., Ries, C., Bandes, R.M., Franklin, B.: Hack the Stack: Using Snort and Ethereal to Master the 8 Layers of an Insecure Network. Elsevier (2006)

9. Herzog, A., Shahmehri, N., Duma, C.: An ontology of information security. Int. J. Inf. Secur. Priv. (IJISP) **1**(4), 1–23 (2007)

10. Joshi, A., Lal, R., Finin, T., Joshi, A.: Extracting cybersecurity related linked data from text. In: 2013 IEEE Seventh International Conference on Semantic Computing, pp. 252–259. IEEE (2013)

11. Jouini, M., Rabai, L.B.A., Aissa, A.B.: Classification of security threats in information systems. Procedia Comput. Sci. **32**, 489–496 (2014)

12. Kalinin, N.: Towards ontology-based cyber threat response (2020)

13. Lamy, J.B.: Owlready: ontology-oriented programming in python with automatic classification and high level constructs for biomedical ontologies. Artif. Intell. Med. **80**, 11–28 (2017)

14. Lamy, J.B., Venot, A., Duclos, C.: Pymedtermino: an open-source generic API for advanced terminology services. In: Digital Healthcare Empowering Europeans, pp. 924–928. IOS Press (2015)

15. Martin, R.A., Barnum, S.: Common weakness enumeration (CWE) status update. ACM SIGAda Ada Lett. **28**(1), 88–91 (2008)

16. Martins, B.F., Serrano, L., Reyes, J.F., Panach, J.I., Pastor, O., Rochwerger, B.: Conceptual characterization of cybersecurity ontologies. In: Grabis, J., Bork, D. (eds.) PoEM 2020. LNBIP, vol. 400, pp. 323–338. Springer, Cham (2020). https://doi.org/10.1007/978-3-030-63479-7_22

17. Onwubiko, C.: Cocoa: an ontology for cybersecurity operations centre analysis process. In: 2018 International Conference On Cyber Situational Awareness, Data Analytics And Assessment (Cyber SA), pp. 1–8. IEEE (2018)

18. Rose, S., Borchert, O., Mitchell, S., Connelly, S.: Zero trust architecture. Technical report, National Institute of Standards and Technology (2019)

19. Salem, M.B., Stolfo, S.J.: Modeling user search behavior for masquerade detection. In: Sommer, R., Balzarotti, D., Maier, G. (eds.) RAID 2011. LNCS, vol. 6961, pp. 181–200. Springer, Heidelberg (2011). https://doi.org/10.1007/978-3-642-23644-0_10

20. Sayan, C.: Automated cyber vulnerability analysis using machine reasoning (2020)

21. Scarpato, N., Cilia, N.D., Romano, M.: Reachability matrix ontology: a cybersecurity ontology. Appl. Artif. Intell. **33**(7), 643–655 (2019)

22. Sokolov, I., et al.: Modern EU research projects and the digital security ontology of Europe. Int. J. Open Inf. Technol. **6**(4), 72–79 (2018)

23. Strom, B.E., Applebaum, A., Miller, D.P., Nickels, K.C., Pennington, A.G., Thomas, C.B.: MITRE ATT&CK: Design and philosophy. Technical report (2018)

24. Sun, J., Xing, Z., Guo, H., Ye, D., Li, X., Xu, X., Zhu, L.: Generating informative CVE description from ExploitDB posts by extractive summarization. arXiv preprint arXiv:2101.01431 (2021)
25. Syed, R.: Cybersecurity vulnerability management: a conceptual ontology and cyber intelligence alert system. Inf. Manage. **57**(6), 103334 (2020)
26. Syed, Z., Padia, A., Finin, T., Mathews, L., Joshi, A.: Uco: a unified cybersecurity ontology. In: Workshops at the Thirtieth AAAI Conference on Artificial Intelligence (2016)
27. Tao, Y., Li, M., Hu, W.: Research on knowledge graph model for cybersecurity logs based on ontology and classified protection. J. Phys. Conf. Ser. **1575**, 012018 (2020)
28. Undercoffer, J., Joshi, A., Pinkston, J.: Modeling computer attacks: an ontology for intrusion detection. In: Vigna, G., Kruegel, C., Jonsson, E. (eds.) RAID 2003. LNCS, vol. 2820, pp. 113–135. Springer, Heidelberg (2003). https://doi.org/10.1007/978-3-540-45248-5_7
29. Välja, M., Heiding, F., Franke, U., Lagerström, R.: Automating threat modeling using an ontology framework. Cybersecurity **3**(1), 1–20 (2020). https://doi.org/10.1186/s42400-020-00060-8
30. Zhang, K., Liu, J.: Review on the application of knowledge graph in cyber security assessment. In: IOP Conference Series: Materials Science and Engineering. vol. 768, p. 052103. IOP Publishing (2020)

Social Network Analysis of the Professional Community Interaction—Movie Industry Case

Ilia Karpov(✉) and Roman Marakulin

Higher School of Economics, Moscow, Russia
karpovilia@gmail.com

Abstract. With the rise of the competition in the movie production market, because of new players such as Netflix, Hulu, HBO Max, and Amazon Prime, whose primary goal is producing a large amount of exclusive content in order to gain a competitive advantage, it is extremely important to minimize the number of unsuccessful titles. This paper focuses on new approaches to predict film success, based on the movie industry community structure, and highlights the role of the casting director in movie success. Based on publicly available data we create an "actor"-"casting director"-"talent agent" - "director" communication graph and show that usage of additional knowledge leads to better movie rating prediction.

Keywords: Social network analysis · Movie rating · Machine learning · Prediction · IMDb · Node2Vec

1 Introduction

The movie industry is undergoing significant changes. Online streaming platforms such as Netflix, HBO Max, Hulu, Disney Plus, Amazon Prime and others, have become extremely popular. The main goal of these platforms is to get as many recurrent users as possible by creating exclusive titles that are only available on one of the platforms, and the amount of content will increase rapidly over the next few years and competition between movie producers will become more fierce.

In order to produce greater amounts of movie titles, producers must invest more in production, considering that it is an extremely risky and expensive type of investment. Media companies or streaming platforms are eager to minimize the number of unsuccessful titles because they will lose their competitive advantage, and lose their market share. This leads to new models and approaches to the estimation of future success of a movie based on the data that is available before a company invests in a project

Traditional solutions classify a future title's success based on statistical information like budget, genre, duration, and so on. The movie industry seems to be

© Springer Nature Switzerland AG 2022
A. Pozanenko et al. (Eds.): DAMDID/RCDL 2021, CCIS 1620, pp. 36–50, 2022.
https://doi.org/10.1007/978-3-031-12285-9_3

quite a close community where a lot of people interact with each other by taking part in similar projects or having the same talent agents - the people who find jobs for an actor and process the incoming offers. To the best of our knowledge, there are no specific approaches to predict movie success when taking into account the information about title principals and their position in the movie industry community. Those features seem to be important because it may be useful to have a casting director who has authority in the movie community and will more likely be able to gather the best cast that will lead a movie to success. However, success may be considered in different ways: academy award, earnings, and rating. In this study, success will be measured by movie ratings taken from IMDb, because awards are often given to movies that meet the current public agenda. Earning strongly depends on marketing spending and brand popularity, for example "The Avengers: End Game" had a $200 million marketing budget with an extremely popular brand and 220 million production budget that led to $1.6 billion in box office revenue.

The main contributions of the paper are the following:

- We propose a joint dataset that incorporates statistical and social network data from three different sources. To the best of our knowledge, this is the first dataset that takes into account social contacts of casting directors and talent agents. The dataset and project information can be found at the github repository[1].
- We study the network structure of the movie community and use the obtained information to predict the future success of a movie title. Our experiment shows that social network information can improve classification accuracy from 4 to 6 % depending on the classification method
- We find out that a casting director is more important than actors in terms of feature importance for the film rating. This leads to the opportunity of predicting the movie's success at an early stage when actors are not even approved for the role.

The rest of this paper proceeds as follows. Section 2 contains an overview of the previous studies regarding movie community analysis and a titles' rating prediction. Section 3 describes the obtained dataset. The 4-th Section contains the details of the movie community graph generation and description. Section 5 is devoted to a movie success classification model, and the conclusion is expressed in 6 Section.

2   Related Work

In this chapter, previous studies related to graph-based movies analysis and its success prediction are observed.

Michael T. Lash and Kang Zhao [9] were focused on feature engineering that allowed them to predict the success of a movie more accurately. They built a

---

[1] https://github.com/karpovilia/cinema.

model that classified movies into 2 groups depending on ROI. As a result, they have achieved a ROC-AUC 0.8–0.9. The data sample contained 14,097 films with 4,420 actors. The independent variables were mainly monetary, for example, the average profit of films for the period, how much the actors earn on average from a film, etc. The social network analysis was used to extract indications of the interactions between actors and directors and use it as features in a classification model.

The authors of the paper "We Don't Need Another Hero - Implications from Network Structure and Resource Commitment for Movie Performance", have made a similar work [11]. They used common movie attributes such as year, budget, and etc. Also, they used SNA methods to extract features about the interactions between actors in movies, for example, the normalized number of contacts of a given film's crew with teams from other movies.

Krushikanth R. Apala and others [3] collected data from social networks, such as comments on trailers on YouTube, the popularity of actors on their Twitter pages and so on, and predicted the success of the film in terms of money. As the main result, they showed that the popularity of the actors is a very important factor in the success of a film.

Bristi, Warda Ruheen, Zakia Zaman, and Nishat Sultana [5] have been trying to improve IMDb rating prediction models using five different machine learning approaches: bagging and random forest classifiers, decision tree classifier, KNearest Neighbours and Naïve Bayes classifiers. Also, the researchers divided movie titles into classes based on their IMDb rating (Fig. 5), making it a dependent variable for classifiers. In order to handle class imbalance, authors used the Synthetic Minority Oversampling Technique (SMOTE) algorithm which balances classes by taking random observations from datasets that are close in feature space, and generating new samples by linear interpolation between selected data records. Also, the researchers used resampling. As a result, they achieved a 99.23 accuracy metric using random forest. However, there were only 274 movie titles in the dataset and the results were provided for a training dataset only.

Another attempt to predict IMDb scores was taken by Rudy Aditya Abarja and Antoni Wibowo in their article "Movie Rating Prediction using Convolutional Neural Network based on Historical Values" [2]. The main idea of the article is to implement Convolutional Neural Network (CNN) to predict a movie's IMDb rating. Authors utilized the IMDb movie data from Kaggle which contained historical features, such as average movies rating by director, an actor's average movie rating, average genre rating and so on. Also, some metadata features, such as budget, duration and release date, were used. As a result, the researchers achieved 0.83 MAE, which is 0.11 better than the best baseline model with 0.94 MAE. Also, Ning, X., Yac, L., Wang, X., Benatallah, B., Dong, M., & Zhang, S showed [12] that deep learning models outperform classical machine learning baselines on the IMDb datasets.

3   Dataset Exploration

In this paper we use joint features from several publicly available databases. The first part of the dataset was obtained from the official IMDb web page which provides the following information: title name, language, production year, duration in minutes, genre, director name, writer name, film crew, including persons primary profession, name, and date of birth, IMDb rating, and number of votes. The second part of the dataset is from the Rotten Tomatoes database that adds information about a movie statistic, such as tomatometer rating, number of votes, number of positive votes, and number of negative votes. This information is used as the retrospective data about a film crew's previous projects. The third part of the dataset is IMDb Pro - an extension to IMDb that contains more complete information about people in the movie industry, for example, talent agent contacts, filmography with current status for every movie, his earning from a movie title, extended biography, and some ratings such a STARmeter and number of news articles, which shows how popular an actor is. The data about his talent agent's contact allows to make a connection between an actor and a talent agent. The combined dataset contains information about 85,855 movie titles.

Based on this data the following features were extracted:

- Country in which a movie has been produced;
- Genre of a movie title, for example, horror or drama;
- Title, that is the name of the title and its IMDb ID;
- Type of production, for example, movie or series;
- Year a movie title has been produced;
- Duration, that is a movie title runtime in minutes;
- Primary language of the movie/series;
- Movie crew, that is actors' names and their IMDb IDs, director name and IMDb ID
- IMDb user rating;
- Number of votes;
- Metascore user rating;
- Number of reviews from critics and users on IMDb;
- Budget of a movie that has been spent on production;
- Income, that is gross income worldwide;
- User rating on RottenTomatoes;
- Age rating (PG, PG-13 and etc.)

Most of the budget, USA gross income, metascore, and worldwide gross income values are missing.

IMDb score distribution is left skewed but quite close to normal 1 There is no need to try to make the distribution normal, because in this research the classification problem is considered, so the IMDb scores will be divided into the following four intervals: 0–3, 4–5, 6–7, 8–10. Rotten Tomatoes users more often rate films with a lower number of points (Fig. 2), making the distribution quite close to the IMDb rating distribution.

As can be seen in Fig. 3, the IMDb score does not heavily depend on genre. Only adult, horror, and sci-fi movies have significantly lower average IMDb scores, however, these genres make a small portion, around 1%, of all records. Also, it can be noted that documentary movies have a better average IMDb

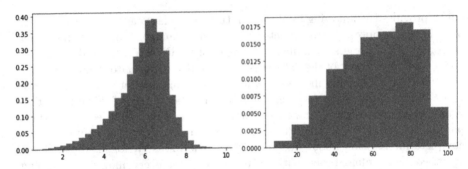

Fig. 1. IMDb rating distribution

Fig. 2. RottenTomatoes user rating distribution

rating, which is probably because these movies are usually filmed specifically for smaller groups of people, for example, the National Geographic fans, and they are not often presented in cinemas or featured on Netflix and other streaming platforms. Because of this, they hold a small portion of observations in the dataset.

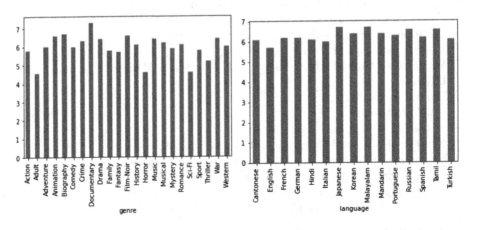

Fig. 3. IMDb rating among genres

Fig. 4. IMDb rating among languages

Language also does not have a significant influence on the IMDb score (Fig. 4), which is quite obvious and leads to the intuition that it does not matter where a movie was produced, and if it is good, it gets high rating, if it is bad, it gets low rating. Most of the movie titles have English, French, Spanish and Japanese as primary languages.

4   Network Model

4.1   Graph Generation

In order to analyze the movie industry community, it should be presented in the form of a directed graph by the following rules:

- Actors, directors, casting directors and talent agents are nodes;
- Actor - actor link appears when two actors mutually take part in a movie title, the link is bidirectional;
- Actor - director link appears when an actor and director mutually take part in a movie title as actor and director respectively,. The direction of the link comes from director to actor, because a director approves an actor for the role in a movie;
- Actor - talent agent link appears when an actor is the client of a talent agent,. The direction is from actor to talent agent because agents work for actors, assigning them to roles and castings;
- Actor - casting director link appears when an actor and casting director mutually take part in a title, as actor and casting director respectively. The direction is from casting director to actor because casting directors find actors who would fit a role;
- Director - casting director link appears when a director and casting director mutually take part in a movie title as director and as casting director, respectively. The direction is from director to casting director because director approval is required.

Following this approach, the entire graph has 50 million links for the existing dataset, so, due to computational limitations, the number of nodes was decreased using the Forest-fire based [10] algorithm. In the first step, the one hundred most popular actors have been taken. Following this, titles, where an actor has starred, should be taken, then the rest of the cast, directors and casting directors should be taken, This should be repeated until one huge connected component is created. Finally, talent agents should be added to the graph. The generated graph has 59,443 nodes and 704,175 edges. The average clustering coefficient is 0.57, and the average shortest path is 8.

4.2   Graph Description

The graph (Fig. 5) has a core-periphery power law structure. The top three nodes by degree are casting directors: Mary Vernieu (casting director for "Deadpool" and another 421 titles), Kerry Barden (casting director for "The Spotlight"), Avy Kaufman (casting director for "The Sixth Sense"). The actor with the highest degree is Nicolas Cage.

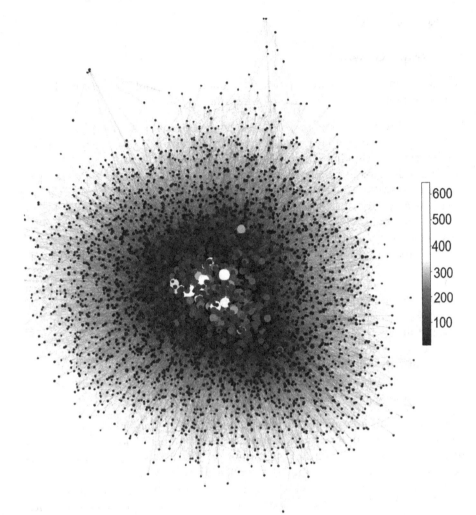

Fig. 5. Movie industry community graph

Figure 6 shows that the graph follows the power-law distribution of the node degrees - most of the nodes have a small degree, which stands for new or unknown actors mostly. Most of the nodes have small eigenvalue centralities which can as nodes with high level of influence are not connected directly. Betweenness centrality distribution (Fig. 7) is concentrated in the area of small values as well - there is no nodes that strongly control the network, and shortest paths often go through different nodes. Hubs and authorities algorithm shows that the top hubs are casting directors and the top authorities are highly popular actors, such as Nicole Kidman, Robert De Niro, Bruce Willice, Julianne Moore and Morgan

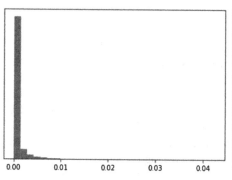

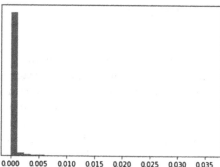

Fig. 6. Degree centrality distribution

Fig. 7. Betweenness centrality distribution

Freeman. The results seem to be reasonable, because casting directors indeed work with a lot of people in movie industry and have a lot of contacts.

4.3 Random Walk Network Model

Node2Vec is an algorithm that generates nodes representations in vector format. The algorithm creates low-dimensional nodes representations of a graph using random walks through a graph. The main idea of the method is that random walks through a graph may be considered as a sentence in a corpus in the following way [8]:

- Every node is treated like a single word;
- Every walk through nodes of a graph is a sentence;
- Then, a skip-gram may be applied to every sentence, that was created by a random walk;

The main parameters of Node2Vec algorithm that may be controlled and adjusted are number of walks, length of each walk, p - the probability to return to the neighbour of the previous node, q - controls the probability to go away from the previous node during a random walk. In the current study we consider number of walks and its length as constants equal to 200 and 80 respectively, because it seems to be relevant for the graph of this size and computational power limits. Parameters p and q are considered from 1 to 4 each, so there are 16 models for each pair of parameters, since there is no intuitive understanding of the best hyperparameters for the current graph. Further in the research, each model will be used separately in prediction models. As a result, the model returns a 24-dimensional vector, that describes a node, which is a member of the movie industry community.

5   Experiment

First of all, the past average ratings of a movie title's crew should be considered, because these variables were important in the previous studies reviewed in the chapter two. We will calculate average IMDb and RottenTomatoes rating of the movies for a director, casting director, writer and actors, so the they will be considered in the models and we will be able to see whether the additional SNA information about their position in the network will affect the results or not. Secondly, graph embeddings for movie team members should be obtained. In this study, we will consider vectors for directors, casting directors and four main actors of a movie. In case of boosting model the mean vector for actors will be taken and in case of neural network model embedding vectors will be taken as is.

Also, because of reduction of the amount of data considered in chapter three the distribution of among buckets should be the same as in original dataset. In our case the original dataset has 53% of the observations into "6–7" bucket, 37% into "4–5" bucket, 8% into "0–3" bucket and 2% into "8–10" bucket. The reduced dataset has 56% of the observations into "6–7" bucket, 35% into "4–5" bucket, 7% into "0–3" bucket and 2% into "8–10" bucket, so the new distribution among classes is quite similar to the original.

Finally, since the current dataset is imbalanced, we perform SMOTE algorithm [6] in order to get equal number of samples in each class.

As for the comparison metric of the classification results the accuracy will be used [4], because it is simple and suitable for the current multiclass classification problem.

5.1   Random Forest Models

The first model to test whether social network analysis features from the movie industry community does improve the model is gradient boosting over random forest using CatBoost [7] Python library.

The best results for the gradient boosting over random forest classification without SNA parameters were obtained with the following parameters:

- Depth = 5;
- l2_leaf_reg = 1;
- Learning rate = 0.05

The best Accuracy = 0.68.

The best results for the gradient boosting over random forest classification with SNA parameters were obtained with the following parameters:

- Depth = 4;
- l2_leaf_reg = 1;
- Learning rate = 0.03

Table 1. Accuracy results for different P and Q

P,Q	1	2	3	4
1	0.7404	0.7368	0.7363	0.7371
2	0.7375	**0.7410**	0.7386	0.7373
3	0.7407	0.7355	0.7371	0.7360
4	0.7353	0.7386	0.7368	0.7393

The best accuracy is 0.74, it is 6% gain due to social network features of the movie industry community.

Comparing the results between models with different features produced by Node2Vec with different parameters, shows no much difference between the results, which is around 0.05 between the most extreme cases, however, the best was obtained with P = 2 and Q = 2 (Table 1).

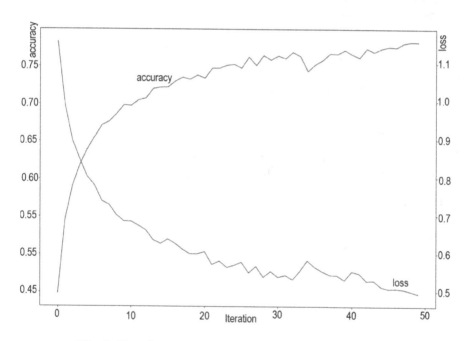

Fig. 8. Neural network model coverage with graph features

According to the feature importance, the most important network features were the ones describing directors' nodes, the second were the ones describing casting directors' and third were the ones describing actors' nodes. Basically, that result shows the importance of casting director who is always in the background, because the importance of each feature indicates how much on average the prediction changes if the feature value changes.

5.2   Random Forest Model

The second test to check whether the SNA features affects the prediction results or not is to use random forest (RF) classification model. To find the optimal parameters grid search approach was used.

The best results for the RF model without SNA parameters were achieved with the following set of the parameters:

– Depth = 6;
– max_features = auto;
– n_estimators = 250;

The best results for the RF model with SNA parameters were achieved with the following set of the parameters:

– Depth = 5;
– max_features = auto;
– n_estimators = 350;

The best accuracy result for the model without SNA features is 0.638, and the best accuracy for the model with SNA features is 0.671. So, as we can see, the social network features improved the result, but it is worse than the best accuracy score of the previous model.

5.3   Decision Tree Model

The third model is a simple decision tree. Once again, optimal parameters were chosen by the grid search approach.

The best results for the model without SNA parameters were achieved with the following set of the parameters:

– max_depth = 300;
– max_features = auto;
– criterion = gini;

The best results for the model with SNA parameters were achieved with the following set of the parameters:

– Depth = 400;
– max_features = auto;
– criterion = gini;

The best accuracy result without SNA is 0.611, and the best result with SNA is 0.601. So, in this case, the results are quite similar, that is probably because of the high input dimension.

Table 2. Neural network architecture

Layer type	Shape	Param
Dense	64	18688
Dropout	64	0
Dense	16	1040
Dropout	16	0
Dense	4	68

5.4 Neural Network Models

The final test is to build a simple neural network and compare the results considering different input vector for the actors, taking it as is. The classification model without movie industry community network features [1] architecture contains three layers with dropout:

The result for the first model is 0.74, which is better comparing to the relative result obtained with the gradient boosting over the random forest model.

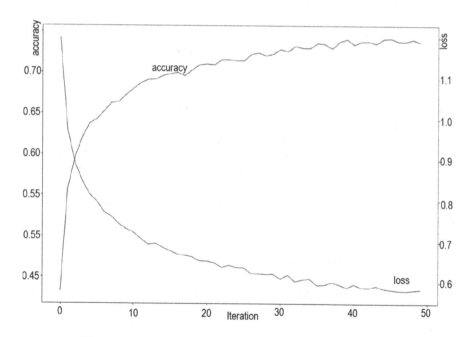

Fig. 9. Neural network model coverage with graph features

The same model with additional movie industry community network features has achieved accuracy = 0.78 (Fig. 9), which is better than the model without these features and better than boosting model created earlier. However, the difference between neural network models with and without graph features has become slightly smaller.

6   Conclusions

Features extracted from the movie industry community graph do improve predictions results as shown in Table 3. A director of a movie has more influence on movie success than a casting director and actors. Also, we can notice that there is no much difference between Node2Vec model that have different random walk parameters.

Table 3. Accuracy results for different models

	Without SNA features	With SNA features
RF + GB	0.684	0.743
RF	0.638	0.671
Decision Tree	0.611	0.601
NN	0.741	**0.781**

Also, as it can be seen in the 3 most of the models showed that social network features obtained using Node2Vec to get the embeddings improved the prediction results even though data without SNA has information about movie crew - average rating for the previous titles, so it can be concluded that SNA features are significant and a position in the movie industry community does affect future movie success.

In the present research the movie industry community have been studied by actor-casting director-talent agent-direct graph analysis, and the following goals have been completed:

- The unique data about casting directors and talent agents has been collected by scrapping web pages;
- The network-graph structure of the movie community has been presented and analyzed, providing information that describes how this community works;
- The embeddings that describe nodes have been obtained;
- The models predicting IMDb score has been created and the information of the different combinations of the feature sets has been collected and analyzed.

Based on the obtained results by achieving these goals it may be concluded that movie community industry does follow the common social network properties, such as power-law distribution, that is there are small number of members who has a significant number of connections with other people from the community. Also, it has been shown that casting directors are important members of the movie community as they are network hubs. Moreover, during the current research is has been shown that features that describes network structure of the movie industry community influence the IMDb rating. Group of features regarding director of movie titles is the most important for the classification, the second most important is the group of features that contains information about casting directing directors, so, these people do affect the movie success. Also, as it was described in the previous studies, the network model shows better results predicting IMDb score.

However, this study has some limitations, such as reduced dataset due to computational resources, so the future researches may include:

- Analysis of the whole graph with all principals of a movie, because it may give more insights;
- Usage of the social network model for best strategies in order to get the cast, because it may be more efficient to spend resources on a popular casting director or actor or talent agent that will bring other people because they are valuable members of the community

Acknowledgements. The article was prepared within the framework of the HSE University Basic Research Program.

References

1. Abadi, M., et al.: Tensorflow: a system for large-scale machine learning. In: Proceedings of the 12th USENIX Conference on Operating Systems Design and Implementation, p. 265–283. OSDI 2016, USENIX Association, USA (2016)
2. Abarja, R.A., Wibowo, A.: Movie rating prediction using convolutional neural network based on historical values. Int. J. Emerg. Trends Eng. Res. 8(5), 2156–2164 (2020). https://doi.org/10.30534/ijeter/2020/109852020
3. Apala, K.R., Jose, M., Motnam, S., Chan, C.C., Liszka, K., Gregorio, F.D.: Prediction of movies box office performance using social media. In: 2013 IEEE/ACM International Conference on Advances in Social Networks Analysis and Mining (ASONAM 2013), pp. 1209–1214 (2013)
4. Baldi, P., Brunak, S., Chauvin, Y., Andersen, C.A., Nielsen, H.: Assessing the accuracy of prediction algorithms for classification: an overview. Bioinformatics 16(5), 412–24 (2000)
5. Bristi, W.R., Zaman, Z., Sultana, N.: Predicting IMDb rating of movies by machine learning techniques. In: 2019 10th International Conference on Computing, Communication and Networking Technologies (ICCCNT), pp. 1–5 (2019)
6. Chawla, N., Bowyer, K., Hall, L., Kegelmeyer, W.P.: Smote: synthetic minority over-sampling technique. J. Artif. Intell. Res. 16, 321–357 (2002)
7. Dorogush, A.V., Ershov, V., Gulin, A.: Catboost: gradient boosting with categorical features support. ArXiv abs/1810.11363 (2018)
8. Grover, A., Leskovec, J.: node2vec: scalable feature learning for networks. In: Proceedings of the 22nd ACM SIGKDD International Conference on Knowledge Discovery and Data Mining (2016)

9. Lash, M.T., Zhao, K.: Early predictions of movie success: the who, what, and when of profitability. J. Manag. Inf. Syst. **33**, 874–903 (2016)

10. Leskovec, J., Kleinberg, J., Faloutsos, C.: Graph evolution: Densification and shrinking diameters. ACM Trans. Knowl. Discov. Data **1**(1), 2-es (2007). https://doi.org/10.1145/1217299.1217301

11. Meiseberg, B., Ehrmann, T., Dormann, J.: We don't need another hero—implications from network structure and resource commitment for movie performance. Schmalenbach Bus. Rev. **60**, 74–98 (2007). https://doi.org/10.1007/BF03396760

12. Ning, X., Yac, L., Wang, X., Benatallah, B., Dong, M., Zhang, S.: Rating prediction via generative convolutional neural networks based regression. Pattern Recogn. Lett. **132**, 12–20 (2020). https://doi.org/10.1016/j.patrec.2018.07.028, https://www.sciencedirect.com/science/article/pii/S0167865518303325. Multiple-Task Learning for Big Data (MTL4BD)

Data Analysis in Astronomy

Cross-Matching of Large Sky Surveys and Study of Astronomical Objects Apparent in Ultraviolet Band Only

Aleksandra S. Avdeeva[1,2](✉), Sergey V. Karpov[3,4,5], Oleg Yu. Malkov[1], and Gang Zhao[6]

[1] Institute of Astronomy, 48 Pyatnitskaya St., 119017 Moscow, Russia
avdeeva@inasan.ru
[2] HSE University, 20 Myasnitskaya Ulitsa, 101000 Moscow, Russia
[3] Institute of Physics, Czech Academy of Sciences, 182 21 Prague 8, Czech Republic
[4] Special Astrophysical Observatory, Russian Academy of Sciences,
369167 Nizhnij Arkhyz, Russia
[5] Kazan Federal University, 420008 Kazan, Russia
[6] Key Laboratory of Optical Astronomy, National Astronomical Observatories,
Chinese Academy of Sciences, 100012 Beijing, China

Abstract. Cross-matching of various information sources is a powerful tool in astronomy that helps to not only enrich and augment the contents of individual ones, but also to discover new and unique objects. In astronomy, cross-matching of catalogues is a standard tool for getting broader information on the objects by combining their data from the surveys performed at different wavelengths, and it allows to solve number of tasks like studying various populations of astronomical objects or investigating the properties of interstellar medium. However, the analysis of objects present in just one catalogue and missing the counterparts in all others is also a promising method that may lead to the discovery of both transients and objects with extreme color characteristics. Here we report on our preliminary search for objects that manifest only in ultraviolet observations by comparing the data from GALEX catalogue with several other surveys in different wavelength ranges. We describe the selection of representative sky surveys for this task and give the details on the process of their cross-matching and filtering of the results. We also discuss the possible nature of several outstanding objects detected during the analysis. We have been able to detect some of faint UV-luminous-only candidates, and a single transient event corresponding to the flare on a cool sdM subdwarf. Additionally, we discuss the potential output of a larger-scale investigation we are planning based on the experience gained during this initial study.

Keywords: Surveys · Stars

A. Pozanenko et al. (Eds.): DAMDID/RCDL 2021, CCIS 1620, pp. 53–73, 2022.
https://doi.org/10.1007/978-3-031-12285-9_4

1   Introduction

The problem of parameterization of astronomical objects based on their photometry is a topical issue. To parameterize a star means to find some of its parameters required for a particular task. A great variety of photometric systems and recently constructed large photometric surveys as well as an emergence of dedicated Virtual Observatory (VO) tools for cross-matching their objects provide an unique possibility to get multicolour photometric data for millions of objects. This combined photometry can be used for relatively accurate determination of the parameters of galaxies, stars and the interstellar medium. In particular, it was shown by [21] that multicolour photometric data from large modern surveys can be used for parameterization of stars closer than around 4.5 kpc and brighter than $g_{\mathrm{SDSS}} = 19.6^m$, including estimation of parallax and interstellar extinction value.

The objects that have been detected in all the surveys under study represent, naturally, the most favourable and convenient material for the research, since the photometric data for them are most abundantly presented and cover the electromagnetic spectrum from ultraviolet (UV) to infrared (IR). However, objects that are found only in one of the surveys under study, and absent in all others, should also attract close attention.

In the process of studying of interstellar extinction [22], we have cross-matched objects from various sky surveys in several selected sky areas and noticed the presence of a significant amount of objects present in GALEX catalogue only, without any counterparts in optical and infrared surveys. Therefore we decided to make a dedicated study of these objects in order to assess their physical nature.

Here we present the pathfinder search for the UV objects from the GALEX survey that have no apparent optical/IR counterparts, or having extreme UV to optical colours, based on an initial study of a small fraction of the sky.

2   Sky Surveys Overview

An astronomical survey is, in contrast to targeted observations, a way to collect data (images or spectra) without specifying the particular object of study. Surveys allow astronomers to analyse astronomical objects statistically as well as to choose a target for a detailed study.

This work will be dealing with six major sky surveys (GALEX, SDSS, Gaia, PanSTARRS, SkyMapper and WISE), two historical all-sky catalogues (USNO-B1.0 and GSC2.3.2) and the photometric catalogue of The Dark Energy Survey (DES DR1). Here we would like to give an overview on these surveys. Main information on the surveys is summarised in Table 1.

Sloan Digital Sky Survey (SDSS) is one of the first modern sky surveys, which observations are carried out using a 2.5-m wide-angle optical telescope equipped with a set of imagers and spectrographs [2]. The telescope is located at Apache Point Observatory in New Mexico. During the existence of the project (it still operates), it had many scientific goals: from the creation of the detailed three-dimensional map of the Milky Way to exoplanet search. The data collection began in 2000 and become available to the public in the form of data releases (usually referred as DR). DR12 contains information on $\approx$470 million objects, it includes all SDSS data taken through 14 July 2014. Data contains images of the sky in five filters named u, g, r, i, z. The response curves - the dependence of the filter transmittance on the wavelength - for these filters and filters of other systems are shown in Fig. 1. In Table 1 we also provide the information on limiting magnitudes - this value answers the question how bright should be a star to be detected by this survey. In astronomy the bigger the number the fainter the star is. The additional information on SDSS project could be found at https://www.sdss.org/dr12/ website.

Pan-STARRS is for Panoramic Survey Telescope and Rapid Response System [11]. It consist of two 1.8-m wide-angle optical telescopes located at Haleakala Observatory, Hawaii. The first telescope and corresponding survey is referred to as Pan_STARRS1 or PS1. An observations on PS1 began in 2010 and operated until 2014. The science goals of the mission include the precision photometric and astrometric measurements of stars in the Milky Way and the Local Group, the search for Potential Hazardous Objects in Solar System, some cosmological constrains and others. PS1 images the sky in five filters: g, r, i, z, y (limiting magnitudes see in Table 1 and corresponding response curves are shown in Fig. 1.). The web page of the project is https://panstarrs.stsci.edu/.

The SkyMapper survey of the southern sky has been running from 2014 to 2019. The observations were carried out with a 1.35-m wide-angle optical telescope of the Australian National University [19]. A broad variety of goals have been declared, including the Milky Way's structure mapping, discovery of new supernovae (SNe), providing information for the comprehensive study of quasars and others. The photometric system (a set of filters) used for observations consist of 6 filters: u, v, g, r, i, z. The last three ones (r, i, z) are very similar to those in SDSS survey, but u and g are moved apart in comparison to SDSS and in the gap between them narrow v filter is inserted.

Gaia is a survey that is being held by the space observatory of the same name [16]. Mission of the project is to measure precisely the positions of stars both in the Milky Way and other parts of the Local Group, i.e. to make a three-dimensional map. The astrometric measurements are complemented by spectral and photometric measurements of all objects. The spacecraft is located at the second Lagrange point. This point lies on the line through the Sun and the Earth, beyond the last one, and is about 1.5 million km far from the Earth. Gaia consist of three scientific instruments (or three-in-one): tools for astrometry and photometry, and the spectrometer used for measurements of objects radial velocity. The additional information could be found at project's web page: https://www.cosmos.esa.int/web/gaia/.

Another space telescope provided all-sky survey is WISE – Wide Field Infrared Explorer [13,34]. The mission was designed to measure the entire sky in infrared. The telescope was launched in December 2009. The All-sky WISE data became available in the form of Data Release by March 2012, it includes all the data obtained from December to August 2010. WISE explores the sky in four filters: W1(3.4 μm), W2(4.6 μm), W3(12 μm), W4(22 μm) (response curves are shown in Fig. 1). The discoveries highlights include the most luminous galaxy in the universe, revealing millions of black holes and discovering new types of stars.

USNO-B1.0 [27] and GSC2.3.2 [20] are all-sky historical catalogues. Both obtained from scanned photographic plates. USNO-B1.0 is compiled from various sky surveys so the colours (the difference in a pair of magnitudes measured with different filters is reffered to as colour) and epochs are not uniform across the catalogue. GSC2.3.2 is the latest release of the second generation Guide Star Catalog. The catalogue is composed of scanned photographic plates from the Palomar and UK Schmidt surveys, so the catalogue also employs different filters and epochs. The initial goal was to provide the support for Hubble Space telescope observation planning, however, today it is employed by various telescopes and missions.

Dark Energy Survey (DES) was designed to examine the expanse of the Universe. For this purpose it aimed at observing large amounts of supernovae, gravitational lenses and galaxy clusters and measuring the distribution of galaxies across the sky. DES explores the sky in optical and near-infrared filters: g, r, i, z, Y. Observations are conducted with 4-m telescope located at Cerro Tololo Inter-American Observatory, Chile. Data Release 1 [1], which is used in this work, consists of the data obtained from August 2013 to February 2016. This data covers nearly 5000 deg^2 of the southern sky. Additional information is available on https://www.darkenergysurvey.org/ website.

GALEX Survey, the one in which catalogue we will be searching for outstanding sources, observe the sky in two ultraviolet filters: near-UV (NUV) and far-UV (FUV). The goal of the mission was to measure the distances to galaxies and the star formation rate in each galaxy. The observations held on orbit using 0.5-m telescope and started on March 2003. The GUVcat_AIS catalogue [7] is enhanced with various tags, that help to provide scientific investigations, e.g. extended source flag. Catalogue also contains low-resolution UV-spectra.

3   Cross-Matching of Multi-Wavelength Surveys

To cross-match several catalogues means to find the emergence of the same astrophysical object in different catalogues (they obviously could have been observed by different instruments, different time or in other words epoch, in different filters etc.) and to link this information. As the process of cross-matching several all-sky catalogues like GALEX is tricky and computationally intensive (see [9] for a successful example of such work, done with the aim opposite to ours), we decided to start with the analysis of a small subset of it to understand better the potential problems and formulate the exact criteria to be used for a later full-scale investigation.

Thus, we made a grid of 2520 sky fields, with 15 arcmin radius each, covering about 1% of the whole sky and uniformly distributed in Galactic coordinates (5° stepping in both directions).

Table 1. Summary of main parameters of sky surveys. N is the number of objects in millions.

Survey	N, 10^6	Sky coverage	Photometric bands and limiting magnitude
GALEX (GUVcat_AIS)	83	55%	$FUV = 19.9^m$, $NUV = 20.8^m$
SDSS DR12	469	35%	$u = 22^m$ $g = 22.2^m$ $r = 22.2^m$ $i = 21.3^m$ $z = 20.5^m$
Pan-STARRS DR1	1919	North of declination $-30°$	$g = 22.0^m$, $r = 21.8^m$, $i = 21.5^m$, $z = 20.9^m$, $y = 19.7^m$
SkyMapper DR1.1	285	48%	$u = 20.5^m$ $v = 20.5^m$ $g = 21.7^m$ $r = 21.7^m$ $i = 20.7^m$ $z = 19.7^m$
Gaia DR2	1693	All Sky	$G = 20^m$ $BP = 20^m$ $RP = 20^m$
WISE	564	All Sky	$W1 = 16.6^m$ $W2 = 15.6^m$ $W3 = 11.3^m$ $W4 = 8.0^m$
USNO-B1.0	1046	All Sky	down to $V = 21^m$
GSC2.3.2	946	All Sky	down to $V = 23^m$
DES DR1	399	11%	$g = 24.33^m$, $r = 24.08^m$, $i = 23.44^m$, $z = 22.69^m$, $Y = 21.44^m$
2MASS	>500	All sky	$J = 15.8^m$, $H = 15.1^m$, $K_s = 14.3^m$

Table 2. Results of cross-matching of various sky surveys used in the present work and listed in Table 1. N_{fields} is the number of sky fields of our grid covered by the survey, with next column showing the same information as a percentile of all sky fields. N_{stars} is the mean number of stars per field for a survey. $P_{matched}$ is the fraction of GALEX objects (only the ones passing the initial criterion from Sect. 3 are used here) matched with a given survey over all fields covered by it. P_{single} is the same fraction for GALEX objects matched with given survey and not matched with all other surveys.

	N_{fields}		N_{stars} per field	$P_{matched}$ %	P_{single}%
GALEX (GUVcat_AIS)	1965	78%	574.1		
SDSS DR12	1152	46%	9078.4	94.9	25.7
PanSTARRS DR1	2040	81%	8788.8	98.5	35.6
SkyMapper DR1.1	1294	51%	1972.1	46.4	0.6
Gaia DR2	2520	100%	5502.1	36.0	11.9
WISE	2520	100%	3427.1	85.5	5.0
USNO-B1.0	2520	100%	4024.9	98.0	11.4
GSC2.3.2	2520	100%	3615.7	98.0	33.8
DES DR1	495	20%	14477.5	97.7	68.0

For every field, we acquired from VIZIER [28] the lists of objects from the set of catalogues summarized in Table 1 and described in more details in Sect. 2. For that set, we selected the data products of major sky surveys (SDSS,

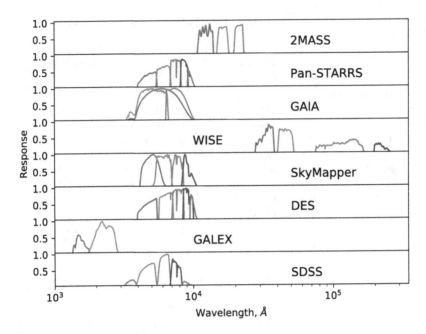

Fig. 1. Response curves of the photometric surveys.

PanSTARRS, SkyMapper, Gaia and WISE), all having uniform coverage of significant fractions of the sky, decent depth (limiting magnitude), uniform photometry in well-defined filters, and providing a representative multi-wavelength coverage. In order to improve the characterization of fainter objects, we included two historical all-sky catalogues (USNO-B1.0 and GSC2.3.2) based on digitization of photographic plates and providing good positional information. Also, in order to improve the coverage of the southern sky, we included the photometric catalogue of DES DR1 covering approximately 5000 square degrees around South Galactic Pole (Fig. 2).

Then, in every field, we cross-matched GALEX objects with all other catalogues using the pairwise distance threshold equal to the hypotenuse of positional accuracy of two surveys. For the latter, we used the conservative value of 5″ for GALEX, 2″ for WISE, 0.1″ for Gaia DR2, and 1″ for all ground-based surveys, the same way as in [21] and [22]. For an initial analysis presented here, we did not perform any comparison of object brightness in various catalogues, thus potentially excluding brighter UV objects where much fainter optical objects are present inside the matching circle.

Quick look study of the results revealed a great number of low-significance GALEX objects not having matched components in other sky surveys (see Table 2). Most of them represent, in our opinion, just a spurious detections[1]. In

---

[1] We leave the thorough investigation of this point and selection of a proper threshold that optimally separates spurious and real detections to a larger-scale follow-up work what will be based on the cross-match of a whole GALEX catalogue.

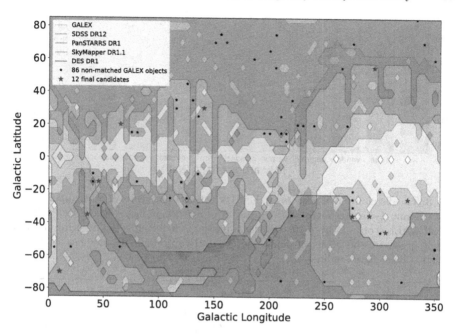

Fig. 2. Sky footprint of GALEX and major optical sky surveys that do not provide all-sky coverage (SDSS DR12, PanSTARRS DR1, SkyMapper DR1.1, DES DR1). All other catalogs used in this study are covering the whole sky, and thus omitted from the plot. The footprint is plotted using small sample fields (0.25° rad) placed on a rectangular grid with 5° step in Galactic coordinates. Overplotted black stars represent the significant (S/N > 3 in both NUV and FUV bands) detections from GALEX catalogue not matched with optical catalogues, while red ones – their final subset passing additional quality cuts as described in Sect. 4. (Color figure online)

order to reliably filter them out, we decided, for this initial analysis, to concentrate just on objects having both NUV and FUV detection with signal to noise (S/N) ratio > 3, i.e. with errors e_NUV < 0.3 and e_FUV < 0.3. After applying such a cut-off, we decreased the total amount of GALEX objects in our sample from 1128082 to 38813, and non-matched ones (we will call them "UV-only") – from 222308 (19.7% of all objects) to 86 (0.2% of the cut-off sample).

4   Properties of UV-Only Objects

Among total of 38813 GALEX objects satisfying our quality criteria (see Sect. 3) in all fields where GALEX data are available, 86 (0.2%) have no counterparts in other catalogues from our list, corresponding to other wavelengths. Hereafter we will call these objects "UV-only objects" to distinguish them from the objects having such counterparts but showing extreme UV to optical colours (we will discuss them later in Sect. 5).

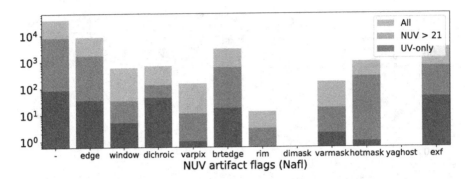

Fig. 3. The distribution of objects by the NUV artefact flags (bits of Nafl, upper panel). Last column also corresponds to the presence of extraction flags (Nexf). Red columns are all GALEX objects from the sky fields passing our initial selection criterion from Sect. 3, green are the subset of fainter (NUV > 21) objects, and blue ones – UV-only objects. (Color figure online)

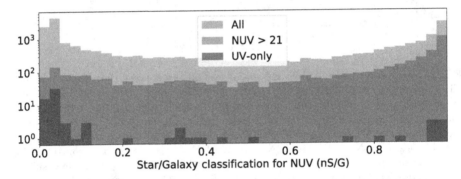

Fig. 4. The distribution of objects by the NUV star/galaxy separation classification probability. Rightmost part corresponds to the highest probability of an object being point-like, while leftmost – to the ones with spatial extent clearly detectable in NUV. Red columns are all GALEX objects from the sky fields passing our initial selection criterion from Sect. 3, green are the subset of fainter (NUV > 21) objects, and blue ones – UV-only objects. (Color figure online)

To exclude various artefacts from selected objects, we checked the quality indicators of the GALEX catalogue [7] – artefact flags (Nafl and Fafl) and extraction flags (Nexf and Fexf). Their distribution is shown for both all sources from our original sample, subset of fainter sources (NUV > 21), and UV-only ones in Fig. 3. There is no striking difference between these three classes of objects except for the slight increase in the frequency of dichroic related (bit 0x04 in Nafl) artifacts for UV-only objects. The latters are among two classes of artifact flags suggested to be excluded in Section 6.2 of [7] (second one – window edge reflections, corresponding to 0x02 bit in Nafl). Top panel of Fig. 5 shows an example of a dichroic artifact among UV-only objects.

ObjID 6378410108349580660

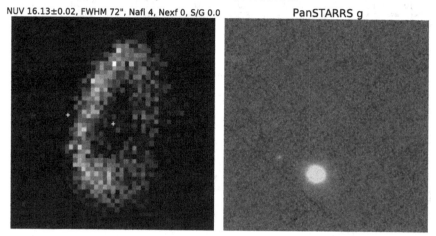

NUV 16.13±0.02, FWHM 72", Nafl 4, Nexf 0, S/G 0.0   PanSTARRS g

ObjID 6374504659153850730

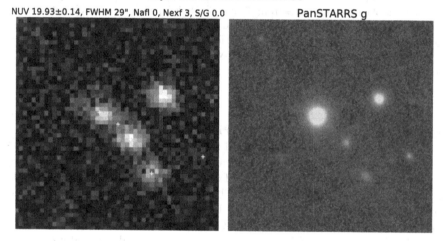

NUV 19.93±0.14, FWHM 29", Nafl 0, Nexf 3, S/G 0.0   PanSTARRS g

Fig. 5. GALEX NUV and Pan-STARRS $72'' \times 72''$ cutouts for some of the UV-only objects, corresponding to the dichroic reflection (top panel, 0x04 bit in Nafl marks this kind of artifacts) and deblending problems (lower panel, 0x02 bit in Nexf marks deblended objects). White crosses denote the positions of GALEX catalogue sources. Both of these kinds of artifacts have low star/galaxy separation score and correspond to the leftmost part of the histogram in Fig. 4. The cutouts are produced using SkyView web service [25].

The presence of extraction flags also represents various types of potential problems, with most prominent being the improper deblending, like the case in lower panel of Fig. 5 where an extra object is erroneously detected between two blended components of a close group. On the other hand, the blend may be

ObjID 6384813626488789397
NUV 20.63±0.12, FWHM 25.2", S/G 0.0

ObjID 6380802643504138553
NUV 20.50±0.12, FWHM 28.8", S/G 0.1

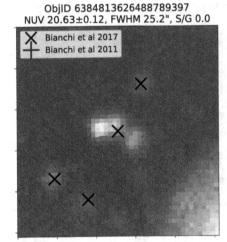
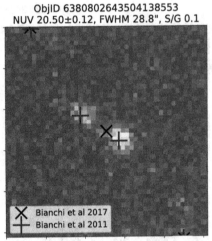

Fig. 6. NUV images of two blended objects from GALEX catalogue. Overplotted are the positions of catalogue entries from [7] and [6]. On left panel, the blend is not resolved in newer catalogue version and not detected at all in older one, while on right panel – again not resolved in newer version but perfectly resolved in older one. In both cases, the star/galaxy separation (nS/G) ratio is very small, clearly indicating an extended object.

unresolved at all, being reported as a single extended object with no artefact or extraction flags set (see an example in Fig. 6).

GALEX photometric pipeline (based on SExtractor code by [4]) also reports the result of a star-galaxy (S/G) classification for all detection based on its fuzziness, corresponding to the probability of an object being point-source (S/G = 1) or extended, galaxy-shaped (S/G = 0). The distribution of these classifications for NUV band is shown in Fig. 4. Of all UV-only objects, only 21 (24%) have a good probability to be point-source (nS/G > 0.5). The rest mostly represent artifacts like the ones discussed in previous paragraph, or truly spatially extended objects like galaxies. Both these classes are unsuitable for our current work, as we are cross-matching with primarily point-source catalogues. Thus, we will concentrate solely on the analysis of point-source objects with nS/G > 0.5. To do that, we add the requirement for the absence of NUV dichroic reflection flags (Nafl&0x04 = 0), and the absence of NUV and FUV extraction flags (Nexf = 0 and Fexf = 0). After such filtering, only 12 objects remain in the final sample which is presented in Table 3.

Among them, none is located inside SDSS DR12 or DES DR1 footprints, while 7 are covered by PanSTARRS DR1 and 9 – by SkyMapper DR1.1. That suggests that the main reason for those objects being undetected may be their faintness – they are below the detection threshold of two latter catalogues (especially SkyMapper which is the least sensitive among them), and at the same time are above the limit of SDSS and especially DES.

Deep co-added PanSTARRS DR1 images are available for 5 of 7 candidates covered by PanSTARRS survey footprint (two more are located below declination of $-30°$, where co-adds are not created despite catalogue data still being available). Visual inspection of four of them (#4, #8, #9 and #10) reveals faint blueish optical uncatalogued sources at the positions of candidates. We performed simple aperture photometry for their fields, calibrated the zero point using PanSTARRS DR1 catalogue entries for several tens of neighbour stars, and estimated the g, r and i magnitudes of these objects. The results, listed in the Comments below Table 3, are marginally brighter than formal depth of PanSTARRS DR1 catalogue listed in Table 2, suggesting that there are some additional selection effects preventing their detection by the official pipeline that explains why these candidates are not matched.

One more candidate, #5, is located on the spiral arm of NGC 4504. Visual inspection of its GALEX data reveals no point source at the object location, however, with a nearby apparent uncatalogued object consistent with $g = 21.5$ PanSTARRS stars at a $8''$ distance from the candidate position. We thus suggest that this candidate is a result of erroneous position measurement, probably due to rapidly varying background due to the nearby galaxy.

For two candidates in PanSTARRS footprint not covered by deep co-added images, there are faint counterparts in an infrared VISTA Kilo-degree INfrared Galaxy (VIKING, [15]) and optical OmegaCAM Kilo-Degree Survey (KiDS, [14]) surveys, with one (#6) being point source, while the other (#12) clearly showing an extended galaxy-like shape and an overall spectral energy distribution (SED) shape typical for active galaxies.

Two more objects (#1 and #2) have an infrared point-like counterparts in VISTA Magellanic Survey (VMC) DR4 [12] survey, with the SED marginally consistent with being a hot massive star.

Finally, three candidates (#3, #7 and #11) are completely undetected on longer wavelengths, being outside of the footprints of deep smaller area southern surveys like VIKING or KiDS.

5   Objects with Extreme UV-optical Colours

Among twelve candidate objects selected in Sect. 4 as "UV-only", i.e. not having counterparts in major optical catalogues listed in Table 2, 8 are actually showing the counterparts at longer wavelengths after a more detailed analysis (with one more most probably being the result of position determination error). Six of them have optical brightness measurements in g, and r filters with sufficient accuracy. Therefore, in order to better assess the parameters of these objects, we decided to study their locations on UV-optical two-colour diagram.

Two-colour diagram is a diagram where the position of an objects is determined by it's pair of colours: the difference between two magnitudes observed in one pair of filters on one axis and the difference between two magnitudes observed in other pair of filters (pairs should not be repeated). Two-colour diagrams are of extreme importance in astronomy, e.g. they allow to obtain the

Table 3. UV-only objects that do not have optical counterparts in Gaia DR2, PanSTARRS DR1, SDSS DR12, SkyMapper DR1.1 and WISE catalogues. ObjID is object identifier according to [7], NUV and FUV are catalogue magnitudes in corresponding energy bands, while Nr and Fr are corresponding source FWHMs (full width at half maximum) in the same bands. Last column refers to the comments given below the table that list the results of additional investigation of these objects.

	ObjID	RA hours	Dec degrees	NUV mag	FUV mag	Nr arcsec	Fr arcsec
#1	6385728416941869657[a]	05 09 37.3843	−64 51 26.1072	20.29 ± 0.15	19.98 ± 0.18	10.8	7.2
#2	6384919152761504385[b]	00 31 04.1513	−72 14 59.604	20.57 ± 0.17	20.76 ± 0.20	18.0	18.0
#3	6383934016112821437	18 49 44.8061	−35 52 48.3816	20.79 ± 0.22	20.28 ± 0.20	14.4	7.2
#4	6375947121772727620[c]	18 28 47.7022	+36 41 12.6384	20.98 ± 0.23	20.82 ± 0.26	18.0	10.8
#5	6382561811642715789[d]	12 32 21.2878	−07 31 39.4356	21.13 ± 0.23	21.38 ± 0.28	21.6	14.4
#6	6380732277981190131[e]	23 23 22.6891	−33 03 40.4208	21.58 ± 0.17	21.23 ± 0.17	18.0	7.2
#7	6387628406320666178	18 48 42.1805	−69 56 56.8356	21.82 ± 0.30	21.13 ± 0.24	14.4	7.2
#8	6379395228892139165[f]	20 06 51.4646	+03 46 09.9516	21.96 ± 0.26	21.08 ± 0.25	10.8	14.4
#9	6379641490505734004[g]	21 02 53.4823	−13 32 22.5888	22.03 ± 0.24	21.16 ± 0.19	14.4	7.2
#10	6373449093631448012[h]	07 51 24.2273	+74 40 12.7524	22.14 ± 0.30	21.92 ± 0.28	10.8	10.8
#11	6385059894544830558	04 16 27.6175	−76 34 41.5488	22.25 ± 0.27	21.12 ± 0.30	7.2	10.8
#12	6380732277981188022[i]	23 22 21.3468	−33 19 27.174	22.72 ± 0.29	22.21 ± 0.27	7.2	14.4

a. This object has a pair of J = 21.9 and J = 22.7 mag point-like counterparts within 3″ in VISTA Magellanic Survey (VMC) DR4 [12].

b. There is also a J = 21.8 point-like source at 4″ in VMC DR4 [12]

c. This object has a faint uncatalogued point-like counterpart visible in PanSTARRS g, r and i band images at a 2″ distance, with rough magnitude estimates of g = 22.45 ± 0.11, r = 22.73 ± 0.18 and i = 23.5 ± 0.3.

d. The object is located on the outer spiral wing of NGC 4504. This object has the lowest (nS/G = 0.75) star/galaxy rating among the sample. Visual inspection of the GALEX NUV cutout do not show any source at the position, but shows an uncatalogued point source at 8″ corresponding to g = 22.5 PanSTARRS star, so we suggest this source to have wrong position in the catalogue.

e. This object has g = 21.78 ± 0.02, r = 22.3 ± 0.03 and i = 22.46 ± 0.12 stellar counterpart in OmegaCAM Kilo-Degree Survey (KiDS) DR3 survey [14], as well as Z=22.4 mag counterpart in VISTA Kilo-degree INfrared Galaxy (VIKING) DR2 survey [15]

f. This object has a faint uncatalogued point-like counterpart visible in PanSTARRS g, r and i band images, with rough magnitude estimates of g = 22.7 ± 0.08, r = 22.8 ± 0.2 and i = 23.4 ± 0.4

g. This object has a faint uncatalogued point-like counterpart visible in PanSTARRS g, r and i band images, with rough magnitude estimates of g = 22.3 ± 0.1, r = 22.8 ± 0.2 and i = 23.9 ± 0.4

h. There is a faint uncatalogued point-like counterpart visible in PanSTARRS g and r band images at a 2″ distance, with rough magnitude estimates of g = 22.6 ± 0.1 and r = 22.8 ± 0.2

i. This object has extended g = 21.76±0.03, r = 21.14±0.02 and i = 20.87±0.05 counterpart in OmegaCAM Kilo-Degree Survey (KiDS) DR3 survey [14], and J = 19.5 mag counterpart in VISTA Kilo-degree INfrared Galaxy (VIKING) DR2 survey [15], with extended emission also clearly visible in the corresponding cutout.

physical parameters of the stars, such as an effective temperature. Stars of different spectral types will be located on different parts of the diagram, so using the color-color diagram one could conclude on spectral type of a star.

To make a two-colour diagram, we selected the deepest catalogue in our sample, DES DR1 [1], and again cross-matched it with GALEX, this time using match radius of 3″, which is an optimal radius according to [9]. In order to exclude artificially large UV-to-optical colours, we selected all ambitious (non-unique) matches and kept only the ones among them with brightest optical components (this way we may bias the colours towards UV deficiency, which is acceptable if we are looking primarily for objects with UV excess). Moreover, we again excluded the GALEX objects with S/N < 3, i.e. e_NUV > 0.3 or e_FUV > 0.3, and the objects with dichroic artifact flags or extraction flags. We also

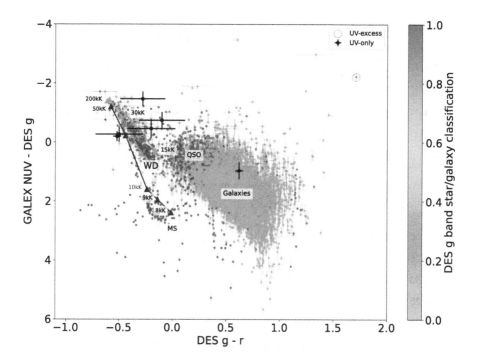

Fig. 7. UV to optical two-color diagram for the objects cross-matched between GALEX and DES DR1 catalogues. Point color corresponds to the DES g band star/galaxy classification rating, with blue corresponding to extended sources, and red – point-like ones. Black circles with error bars correspond to the objects from Table 3 where g and r band photometry is available (see comments there). Red dotted cicrle denotes the only point- source object with (NUV-g) < −2 (see text). Dark violet triangles mark the sequence of white dwarfs of various temperatures and metallicities, while dark red ones – the sequence of hottest main sequence stars, both taken from [9]. Also, the loci of QSOs and galaxies, also taken from the same work, are marked with the text labels. (Color figure online)

excluded DES DR1 objects with g or r band isophotal flags (gIsoFl and rIsoFl) set, and kept only the ones with recommended values of extraction flags (gFlag < 4 and rFlag < 4).

Figure 7 shows the distribution of resulting matches in a two-color DES (g-r) vs (GALEX NUV - DES g) diagram, with color-coded DES g band star/galaxy classification rating clearly separating the loci of stellar (point sources) from extragalactic (extended) objects, with also a well-defined cloud of point-like QSOs in the middle. The plot is mostly analogous to Figures 4 and 5 of [9] and has the same overall layout. The positions of 6 objects from Table 3 where g and r magnitudes are available are shown with black circles with corresponding error bars. The only one (#12) having extended optical emission falls into the extragalactic region, while five others are consistent with the locus of hottest

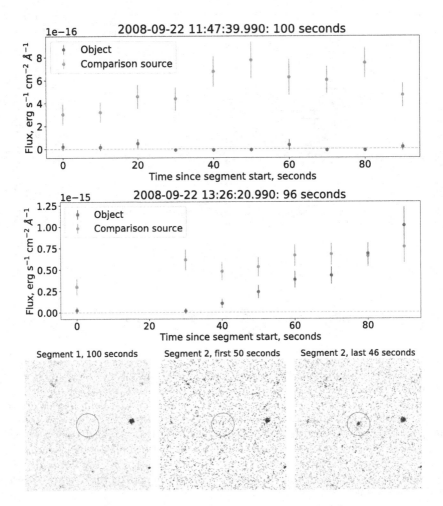

Fig. 8. NUV light curves (two upper panels) of an object with extreme UV-optical colour and a nearby supposedly stable UV source (NUV = 18.27 ± 0.05). The data are acquired and calibrated using *gPhoton* software package by [26]. For that, we chose the 5″ aperture radius, background annulus spanning between radii of 10″ and 60″, and a time step of 10 s (approximately 1/10 of the data segment length). We then excluded from the light curves the points marked as unreliable or problematic by *gPhoton*. The object position was observed twice during the GALEX mission. During the first data segment, its NUV emission is not detectable, while during the second one an apparent flare onset is clearly visible in its light curve. Visual inspection of the time-resolved images (lower panel shows low temporal resolution coadds of first and second data segments, but we also checked higher resolution ones visually) built also with *gPhoton* confirms that it is indeed a flash of the positionally stable source, and not a moving object or some imaging artefact polluting the aperture. Brighter source to the right is the one used for comparison in the light curves. (Color figure online)

stars. Unfortunately, the accuracy of measurements does not allow to pinpoint them to either main sequence or hot white dwarfs tracks.

Anyway, the colors of these objects are quite typical and do not represent any extreme population. Thus we decided to investigate the outliers in the two-colour diagram of Fig. 7 corresponding to the extreme UV excess (we will call them "UV-excess" objects). As a numerical criterion for the latter, we chose (NUV-g) < -2, which exceeds, within typical error bars, the values for the hottest main sequence stars and white dwarfs[2], as well as the locus of extragalactic sources. We also require the objects to be point-like by having S/Gg > 0.5.

After applying these criteria, we got only one source. Positionally coincident with DES J020930.29-533512.6 is GALEX 6384602553542247330, with (NUV-g) $= -2.22 \pm 0.13$, i.e. marginally consistent with our color criterion. The object also has marginal red colours, (g-r) $= 1.70 \pm 0.03$, (r-i) $= 1.87 \pm 0.01$ and (r-z) $= 2.69 \pm 0.01$, placing it well outside the locus of hot stars and suggesting that it is a cool sdM subdwarf [32]. The Gaia EDR3 distance estimate of r $= 337$ pc [3], giving absolute magnitude $M_r = 13.12$ is also consistent with this interpretation.

While strong UV emission is not expected from such cool stars, magnetic activity is quite prominent in them [32]. Thus we decided to investigate the UV emission from this object in more details. For that purpose, we constructed the light curves (the radiation intensity as a function of time) and time resolved images of the object and a nearby brighter UV source using *gPhoton* software package [26] over all two of the time segments (both approximately 100 s long) when the position was observed by GALEX. The light curves are shown in Fig. 8, and it is clearly seen there that the source is essentially invisible (and its flux is consistent with background level) in the first data segment and during the first half of the second one. However, since approximately 40 s into the second segment, the brightness of the source starts to rapidly rise, becoming even brighter than the comparison object (which has NUV $= 18.27 \pm 0.05$) during the last 10 s. Visual inspection of the time resolved images (lower panel of Fig. 8) also confirms that the event is indeed a flare from a point source, and not some moving object or imaging artefact passing the aperture. Thus we may conclude that it is indeed a stellar flare, with timescale and amplitude not unusual for typical flares observed by GALEX [10].

6   Discussion

Our initial search did not reveal any high significance (S/N $\gg 5$) detection of GALEX objects not matched with optical catalogues – UV-only objects. However, several lower significance ones (with S/N around 3 to 5, but having simultaneous detection in both NUV and FUV bands, not marked as an imaging artifacts or blends, and thus most probably corresponding to actual astrophysical

---

[2] The sequences of white dwarfs and main sequence stars of various temperatures are shown in Fig. 7 after [9]. These locations are for unreddened case – however, the reddeding shifts the points there towards bottom-right, and thus cant́ move "normal" object above the limit for hottest stars.

objects) UV-only objects (UVLO candidates) are detected (see Table 3). They are all located outside of sky regions covered by deepest surveys from our sample – SDSS DR12 and DES DR1. However, two of them have faint catalogued counterparts in an infrared VISTA Kilo-degree INfrared Galaxy (VIKING, [15]) and OmegaCAM Kilo-Degree Survey (KiDS, [14]) surveys, while for 4 more we were able to locate and measure faint uncatalogued counterparts in PanSTARRS DR1 deep co-added images. Among these six objects with measured optical colours, five are spanning the locus of hot massive stars in Fig. 7, while one – the locus of galaxies. Two more objects also have infrared point-like counterparts in VISTA Magellanic Survey (VMC) DR4 [12] survey, with the SED also marginally consistent with being a hot massive star. Of the remaining four, one is most probably the result of improper position determination, while final three (#3, #7 and #11) are completely undetected on longer wavelengths, being outside of the footprints of deeper southern sky surveys like VIKING and KiDS.

Our search for GALEX objects that do have optical counterparts but display an unusual – extreme – UV to optical color in the footprint of DES DR1 survey, the deepest among the ones considered in this work, has revealed a single source with the (NUV-g) color exceeding the one for the hottest main sequence stars and white dwarfs. Its detailed study revealed that the object is actually a cool sdM subdwarf, with UV emission apparent during the flare like event, which we attribute to its magnetic activity.

Thus, only one of twelve UV-only objects we selected shows an extended emission on longer wavelengths and an AGN-like SED, while five more occupy the locus of hot stars on a colour-colour diagram (like the ones shown in Figures 4 and 5 of [9]). Indeed, one of the most obvious candidates for ultraviolet luminous objects (UVLO) are hot massive stars. However, even the hottest stars would have (GALEX NUV - SDSS g) color of -1.5 or smaller [8], thus they should also be sufficiently bright in the optical range to be detectable by the modern sky surveys like SDSS or KiDS, while occasionally invisible in the wider field but less sensitive ones like PanSTARRS 3Pi Survey and especially SkyMapper.

One can roughly estimate the maximum luminosity of UVLO candidates in the optical/IR range from the following considerations. The faintest SDSS objects in the high-latitude fields covered by our analysis, where interstellar extinction can be neglected, have $g_{SDSS} \approx 26^m$ and $i_{SDSS} \approx 25^m$. Consequently, UVLO, depending on the distance (d, pc), cannot be brighter than $M_g = 31 - 5 \log d$ and $M_i = 30 - 5 \log d$ (general formula is $M_v = m_v - 5 \log d + 5 - A_v$, where M_v is an absolute magnitude, i.e. the magnitude as it would be seen if the distance for an object was 10pc, m_v is an apparent magnitude, d is a distance to an object measured in pc, A_v is an extinction, v indicates the particular filter). For the characteristic distance of ≈ 250 pc, where the concentration of stars to the Galactic equator starts to appear, it is about $M_g = 19^m$ and $M_i = 18^m$.

For lower galactic latitudes, estimation of the maximum luminosity of UVLO candidates in the optical/IR range should include interstellar extinction. We assume that the distribution of the reddening material quite closely obeys the [30] law of the form

$$A(d, b) = \frac{a_0 \beta}{\sin |b|} \left(1 - e^{-\frac{d \sin |b|}{\beta}}\right),$$ (1)

where a_0 is the value of absorption per kpc in the galactic plane and β is the scale-height. Here we use the average values $a_0 = 1.35 \pm 0.12$ mag per kpc and $\beta = 0.16 \pm 0.02$ kpc obtained for V-band by [29] in their comprehensive study of open clusters. Interstellar extinction in V-band, A_V, calculated with Eq. (1), is transferred to A_λ, using $A_\lambda = R_\lambda * A_V / R_V$, where total-to-selective extinction ratios $R_g = 3.31 \pm 0.03$, $R_i = 1.72 \pm 0.02$ are taken from [35], and R_V is assumed to be 3.1.

Hottest massive stars on the main sequence have $M_V = -6.35$ and thus Galactic ones should be extremely bright unless highly reddened, therefore we may expect only the ones in nearby galaxies to pass our criteria of being non-detectable on longer wavelengths. On the other hand, hot white dwarfs may be as faint as $M_V = 9..12$, and their population in the Solar vicinity (tens of pc up to kpc) may constitute a large fraction of the UVLO. On the more exotic side, strong UV excess is observed in a recent detection of a massive super-Chandrasekhar merger product of binary white dwarf system [18].

Another possible candidate for the role of UVLO could be isolated (single or components of wide binaries) old neutron stars, slowly accreting interstellar matter. This type of objects was proposed by [24] as candidates to the sources detected in the extreme ultraviolet by the ROSAT/WFC and EUVE all-sky surveys and unidentified with any optical counterparts. Such neutron stars are expected to be abundant in the Galaxy, to concentrate towards the Galactic plane, and have not been unambiguously detected yet. Even more abundant, and also still undetected, is the Galactic population of isolated stellar-mass black holes produced as a result of stellar evolution. They may produce faint synchrotron UV and X-ray emission due to spherical accretion of magnetized interstellar matter [5] and thus also contribute to the UVLO population.

It seems useful to compare the observed properties of UVLO candidates with properties of X-ray binaries, as the latter represent a population of compact massive objects. In addition, they emit in the X-ray range, close to the ultraviolet. For comparison, we used objects from the comprehensive list of 137 high-mass and 236 low-mass X-ray binaries [23]. Their spatial distribution is shown in Fig. 9. It can be seen that the distribution of high-mass X-ray binaries (that is, objects that are pairs of initially very massive stars) by galactic longitude is similar to the UVLO distribution.

Finally, we note that our analysis should be able to reveal, apart from the objects with strong UV excesses in the spectra, also the transient ones – the objects what were either only visible during the GALEX observations of their position (and not during all other surveys), or the moving ones corresponding to Solar system bodies. [33] found 1342 detections of 405 asteroids appearing in GALEX images. Several of them fall within our sky fields, and 5 have spatially coincident entries in the revised version of GALEX catalogue [7] we are using. However, none of them has FUV detections (which is consistent with [33]), thus failing our initial quality check. It stresses the necessity of a more sophisticated

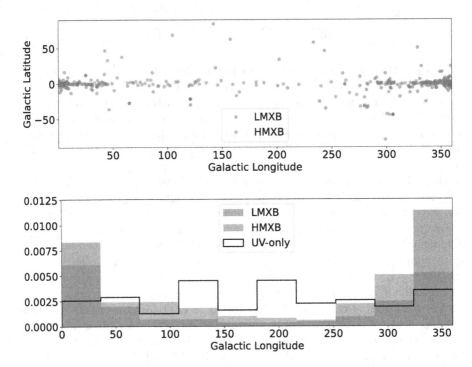

Fig. 9. Scatter of high-mass (HMXB) and low-mass (LMXB) X-ray binaries [23] Galactic coordinates (top panel), and histogram of their Galactic longitudes in comparison with one of UVLO candidates (UV-only objects with no artefact flags and FWHMs consistent with point objects).

treatment of spurious events necessary for a full-scale analysis what we are planning as a continuation of current work.

Transient events detectable by GALEX naturally include the flares on magnetically active stars. The search for such events has been performed e.g. by [10] who have been able to detect 1904 short duration flares on 1021 stars, with amplitudes of flux enhancement reaching 1700 times above quiescent levels. Such events should also appear as either UV-only objects (for distant low-luminosity stars in the Galaxy below the detection limit of deepest optical sky surveys), or sources with extreme UV to optical color (for closer ones, like the flare on a cool sdM subdwarf we detected in Sect. 5).

Other transient sources, which may be detected in the ultraviolet spectral region, are UV-outbreaks prior to SNe. The radius and surface composition of an exploding massive star, as well as the explosion energy per unit mass, can be measured using early UV observations of core-collapse SNe. A theoretical framework to predict the number of early UV SN detection in GALEX and planned ULTRASAT [31] surveys was developed by [17], and the comparison of observations with calculated rates shows a good agreement. Also, it was found by [17] that seven SNe were clearly detected in the GALEX NUV data. Other astrophysical sources potentially yielding a transient UV signal are gamma-ray burst early afterglows, tidal disruption events, and AGN flares.

7 Conclusion

We performed a pathfinder study aimed at the characterization of the objects apparent in UV bands but lacking counterparts at longer wavelengths, or displaying an extreme UV to optical colors – "ultraviolet luminous objects" (UVLO). To do so, we cross-matched the catalogue of GALEX sources with several other modern large scale sky surveys in a number of fields covering about 1% of the sky. Even with quite restrictive quality cuts (which undoubtedly should be relaxed for the follow-up larger scale analysis) we have been able to uncover some of faint UVLO candidates, and a single transient event corresponding to the flare on a cool sdM subdwarf, what testifies the approach we use.

Our initial analysis demonstrated importance of such study, and we plan to continue it using larger-scale cross-matching of the whole GALEX catalogue with other all-sky surveys, combined with a smarter selection criteria and methods for filtering out spurious events in order to reliably uncover the UVLO population, detect individual UVLO objects and perform their detailed investigation.

We hope that the results of the current study allow us to develop a methodology for the detection and investigation of individual UVLO objects and expand it to a large-scale study. Some steps, made here manually, will be automated, while others will remain to require human intervention. At least we can claim to have all the software we need.

Acknowledgement. S.K. was supported by European Structural and Investment Fund and the Czech Ministry of Education, Youth and Sports (Project CoGraDS – CZ.02.1.01/0.0/0.0/15 003/0000437). The work is partially performed according to the Russian Government Program of Competitive Growth of Kazan Federal University. O. M. thanks the CAS President's International Fellowship Initiative (PIFI). Part of this work was supported by the National Natural Science Foundation of China under grant Nos. 11988101, 11890694 and the Chinese-Russian NSFC-RFBR project number 20-52-53009. This research has made use of NASA's Astrophysics Data System, and use of the VizieR catalogue access tool, CDS, Strasbourg, France.

References

1. Abbott, T.M.C., Abdalla, F.B., Allam, S., et al.: The dark energy survey: data release 1. Astrophys. J. Suppl. Ser. **239**(2), 18 (2018). https://doi.org/10.3847/1538-4365/aae9f0
2. Alam, S., Albareti, F.D., Allende Prieto, C., et al.: The eleventh and twelfth data releases of the Sloan digital sky survey: final data from SDSS-III. Astrophys. J. Suppl. Ser. **219**, 12 (2015). https://doi.org/10.1088/0067-0049/219/1/12
3. Bailer-Jones, C.A.L., Rybizki, J., Fouesneau, M., et al.: VizieR Online Data Catalog: Distances to 1.47 billion stars in Gaia EDR3 (Bailer-Jones+, 2021). VizieR Online Data Catalog I/352 (2021)

4. Bertin, E., Arnouts, S.: SExtractor: software for source extraction. Astron. Astrophys. Suppl. **117**, 393–404 (1996). https://doi.org/10.1051/aas:1996164

5. Beskin, G.M., Karpov, S.V.: Low-rate accretion onto isolated stellar-mass black holes. Astron. Astrophys. **440**(1), 223–238 (2005). https://doi.org/10.1051/0004-6361:20040572

6. Bianchi, L., Herald, J., Efremova, B., et al.: GALEX catalogs of UV sources: statistical properties and sample science applications: hot white dwarfs in the Milky Way. Astrophys. Space Sci. **335**(1), 161–169 (2011). https://doi.org/10.1007/s10509-010-0581-x

7. Bianchi, L., Shiao, B., Thilker, D.: Revised catalog of GALEX ultraviolet sources. I. the all-sky survey: GUVcat_AIS. Astrophys. J. Suppl. Ser. **230**, 24 (2017). https://doi.org/10.3847/1538-4365/aa7053

8. Bianchi, L., Efremova, B., Herald, J., et al.: Catalogues of hot white dwarfs in the Milky Way from GALEX's ultraviolet sky surveys: constraining stellar evolution. Mon. Not. R. Astron. Soc. **411**(4), 2770–2791 (2011). https://doi.org/10.1111/j.1365-2966.2010.17890.x

9. Bianchi, L., Shiao, B.: Matched photometric catalogs of GALEX UV sources with Gaia DR2 and SDSS DR14 databases (GUVmatch). Astrophys. J. Suppl. Ser. **250**(2), 36 (2020). https://doi.org/10.3847/1538-4365/aba2d7

10. Brasseur, C.E., Osten, R.A., Fleming, S.W.: Short-duration stellar flares in GALEX data. Astrophys. J. **883**(1), 88 (2019). https://doi.org/10.3847/1538-4357/ab3df8

11. Chambers, K.C., Magnier, E.A., Metcalfe, N., et al.: The Pan-STARRS1 Surveys. arXiv e-prints (2016)

12. Cioni, M.R.L., Clementini, G., Girardi, L., et al.: The VMC survey. I. Strategy and first data. Astron. Astrophys. **527**, A116 (2011). https://doi.org/10.1051/0004-6361/201016137

13. Cutri, R.M. et al.: Vizier online data catalog: Wise all-sky data release (Cutri+ 2012). VizieR Online Data Catalog II/311 (2012). https://ui.adsabs.harvard.edu/abs/2012yCat.2311....0C

14. de Jong, J.T.A., Verdoes Kleijn, G.A., Erben, T., et al.: The third data release of the kilo-degree survey and associated data products. Astron. Astrophys. **604**, A134 (2017). https://doi.org/10.1051/0004-6361/201730747

15. Edge, A., Sutherland, W., Kuijken, K., et al.: The VISTA kilo-degree infrared galaxy (VIKING) survey: bridging the gap between low and high redshift. Messenger **154**, 32–34 (2013)

16. Gaia Collaboration, Brown, A.G.A., Vallenari, A., et al.: Gaia data release 2. Summary of the contents and survey properties. Astron. Astrophys. **616**, A1 (2018). https://doi.org/10.1051/0004-6361/201833051

17. Ganot, N., Gal-Yam, A., Ofek, E.O., et al.: The detection rate of early UV emission from supernovae: a dedicated galex/PTF survey and calibrated theoretical estimates. Astrophys. J. **820**(1), 57 (2016). https://doi.org/10.3847/0004-637X/820/1/57

18. Gvaramadze, V.V., Gräfener, G., Langer, N., et al.: A massive white-DWARF merger product before final collapse. Nature **569**(7758), 684–687 (2019). https://doi.org/10.1038/s41586-019-1216-1

19. Keller, S.C., Schmidt, B.P., Bessell, M.S., et al.: The SkyMapper telescope and the southern sky survey. Publ. Astron. Soc. Aust. **24**(1), 1–12 (2007)

20. Lasker, B.M., Lattanzi, M.G., McLean, B.J., et al.: The second-generation guide star catalog: description and properties. Astrophys. J. **136**(2), 735–766 (2008). https://doi.org/10.1088/0004-6256/136/2/735

21. Malkov, O., Karpov, S., Kovaleva, D., et al.: Verification of photometric parallaxes with Gaia DR2 data. Galaxies **7**(1), 7 (2018)
22. Malkov, O., et al.: Modern astronomical surveys for parameterization of stars and interstellar medium. In: Elizarov, A., Novikov, B., Stupnikov, S. (eds.) DAMDID/RCDL 2019. CCIS, vol. 1223, pp. 108–123. Springer, Cham (2020). https://doi.org/10.1007/978-3-030-51913-1_8
23. Malkov, O.Y., Tessema, S.B., Kniazev, A.Y.: Binary star database: binaries discovered in non-optical bands. Balt. Astron. **24**, 395–402 (2015). https://doi.org/10.1515/astro-2017-0241
24. Maoz, D., Ofek, E.O., Shemi, A.: Evidence for a new class of extreme ultraviolet sources. Mon. Not. R. Astron. Soc. **287**(2), 293–306 (1997). https://doi.org/10.1093/mnras/287.2.293
25. McGlynn, T., Scollick, K., White, N.: SKYVIEW: the multi-wavelength sky on the internet. In: McLean, B.J., Golombek, D.A., Hayes, J.J.E., et al. (eds.) New Horizons from Multi-Wavelength Sky Surveys. Proceedings of the 179th Symposium of the International Astronomical Union, vol. 179, p. 465 (1998)
26. Million, C., Fleming, S.W., Shiao, B., et al.: gPhoton: the GALEX photon data archive. Astrophys. J. **833**(2), 292 (2016). https://doi.org/10.3847/1538-4357/833/2/292
27. Monet, D.G., Levine, S.E., Canzian, B., et al.: The USNO-B catalog. Astrophys. J. **125**(2), 984–993 (2003)
28. Ochsenbein, F., Bauer, P., Marcout, J.: The vizier database of astronomical catalogues. Astron. Astrophys. Suppl. Ser. **143** (2000). https://doi.org/10.1051/aas:2000169
29. Pandey, A.K., Mahra, H.S.: Interstellar extinction and Galactic structure. Mon. Not. R. Astron. Soc. **226**, 635–643 (1987). https://doi.org/10.1093/mnras/226.3.635
30. Parenago, P.P.: On interstellar extinction of light. Astron. Zh. **17**, 3 (1940)
31. Sagiv, I., Gal-Yam, A., Ofek, E.O., et al.: Science with a wide-field UV transient explorer. Astrophys. J. **147**(4), 79 (2014). https://doi.org/10.1088/0004-6256/147/4/79
32. Savcheva, A.S., West, A.A., Bochanski, J.J.: A new sample of cool subdwarfs from SDSS: properties and kinematics. Astrophys. J. **794**(2), 145 (2014). https://doi.org/10.1088/0004-637X/794/2/145
33. Waszczak, A., Ofek, E.O., Kulkarni, S.R.: Asteroids in GALEX: near-ultraviolet photometry of the major taxonomic groups. Astrophys. J. **809**(1), 92 (2015). https://doi.org/10.1088/0004-637X/809/1/92
34. Wright, E.L., Eisenhardt, P.R.M., Mainzer, A.K., et al.: The wide-field infrared survey explorer (wise): mission description and initial on-orbit performance. Astrophys. J. **140**(6), 1868–1881 (2010)
35. Yuan, H.B., Liu, X.W., Xiang, M.S.: Empirical extinction coefficients for the GALEX, SDSS, 2MASS and WISE passbands. Mon. Not. R. Astron. Soc. **430**, 2188–2199 (2013). https://doi.org/10.1093/mnras/stt039

The Diversity of Light Curves of Supernovae Associated with Gamma-Ray Bursts

Sergey Belkin[1,2]($\boxtimes$) (iD) and Alexei Pozanenko[1,2] (iD)

[1] Space Research Institute, Russian Academy of Sciences,
Profsoyuznaya ul. 84/32, Moscow 117997, Russia
`astroboy96@mail.ru`
[2] National Research University "Higher School of Economics",
Myasnitskaya ul. 20, Moscow 101000, Russia

Abstract. More than 10000 gamma-ray bursts (GRBs) have been detected since discovery. Long-term observations of about 850 GRB afterglow in optic since 1998 have shown that a core-collapse supernova (SN) accompanies about 50 nearby GRB sources. We have collected about two dozen SNe' multicolor light curves associated with GRBs. The sample is based on published data, obtained during observations of GRB-SN cases by ground-based observatories all around the world including our own observations. A description of the procedure for the extraction of the SN's light curve, its analysis, and phenomenological classification of SNs are presented. We also discuss the current status and problems of investigations of SN associated with GRB.

Keywords: Astrophysics · Gamma-ray burst · Supernova · Optical counterparts

1 Introduction

1.1 Statistical Information

Gamma-ray bursts (GRBs) are among the most energetic powerful events in the Universe with kinetic energies up to 10^{52} erg. They were first detected by the Vela satellite in 1963. 16 short GRBs were observed during Vela satellites operation from 1969 to 1972 [1].

Further, the development of technologies aimed at studying gamma bursts began. And several years later, from 1978 to 1980 (384 days), observations of gamma-ray bursts were carried out by the KONUS experiment on the spacecrafts Venus 11 and Venus 12. According to the results of these observations, 143 gamma-ray bursts were observed [2].

The next step in increasing the statistics of gamma-ray bursts was the launch of the Compton Gamma Ray Observatory into orbit in 1991. Between 1991 and 1996, the Burst and Transient Source Explorer (BATSE) detector installed at the CGRO observatory detected 1637 gamma-ray bursts [3].

© Springer Nature Switzerland AG 2022
A. Pozanenko et al. (Eds.): DAMDID/RCDL 2021, CCIS 1620, pp. 74–91, 2022.
https://doi.org/10.1007/978-3-031-12285-9_5

These observations made it possible, in particular, to confirm that gamma-ray bursts are distributed isotropically over the celestial sphere [4], and also that they occur outside the Milky Way (e.g. [5]).

Around the same time, the idea was put forward that after the main burst of gamma rays, slow decaying radiation should occur at longer wavelengths, later called "afterglow" and formed as a result of the collision of the ejected matter with the interstellar medium [6]. Previous searches for afterglow were unsuccessful due to the difficulty in quick and accurate determination of the source's location. A breakthrough in this direction came when an X-ray detector onboard the BeppoSAX detected the X-ray afterglow from GRB 970228 [7]. Precise localization of the source made it possible to further detect optical [8] and radio [9] afterglow. During its entire operation from 1996 to 2003, BeppoSAX detected 1082 gamma-ray bursts [10] and at the same time marked the beginning of the afterglow era.

When telescopes had been able to observe the afterglow of the gamma-ray burst, they needed to be improved somehow so that earlier stages of the gamma-ray burst could also be observed. For this, it was necessary to reduce the source identification time and make it less than a minute. Swift [11] became such a spacecraft, which proclaimed the beginning of the era of rapid identification. Launched in 2004, it is still in operation today, regularly providing localization areas for many GRBs. To date, Swift has detected 1645 GRBs [12]. 395 GRBs out of this are with a certain redshift.

Four years after the launch of Swift, the Fermi Gamma-ray Space Telescope was launched. Another spacecraft that still reports the coordinates of, in particular, gamma-ray bursts. For about 12 years of operation of Fermi-GBM, 2356 gamma-ray bursts have been detected [13]. Thus, more than 6800 gamma-ray bursts have been detected over the entire period of their study since 1967.

1.2 Physical Description of the Phenomenon

Based on all these observations, the division of gamma-ray bursts into two classes was carried out. The first one is called "short" ($T_{90} \lesssim 2\,s$; [14]) and these GRBs are associated with the merger of a binary system of neutron stars, or a neutron star and a black hole [15]. The second class includes the so-called "long" ($T_{90} \gtrsim 2\,s$; [14]) gamma-ray bursts and they are associated with a type Ic supernova caused by the collapse of an almost naked core of a massive star at a late stage of its evolution [16].

The light curve of "long" GRBs in the optical wavelength range can be divided into four successive parts. First, there is an active phase, during which the central machine still continues to generate energy.

This is followed by the afterglow phase, which is usually the longest one. This phase is described by a power law or a power law with a break. The break is associated with the geometric effect of the collimation of the gamma-ray burst jet ("jet break"). Also, at the afterglow stage, there can be various kinds of local inhomogeneities. It could be such local deviations from the power law as flashes, bumps and wiggles [17].

Further, approximately 7–30 days after the registration of the gamma-ray burst, an excess of emission associated with the flare-up of the supernova, that is, the expansion of its photosphere with a concomitant increase in luminosity, may appear. This is a supernova phase, which looks like an extended bulge on the light curve of a gamma-ray burst against the background of the power law decay of the afterglow luminosity. The sign of a supernova appears when the work of the central machine of the burst has long ended and there is nowhere to take an additional source of energy.

Finally, when the source decays, the flux of its radiation falls below the luminosity level of the host galaxy. This is the last phase of the long GRB light curve.

Determination of the presence of a GRB-SN connection for a specific gamma-ray burst can be done in two ways: photometric and spectroscopic. Photometric confirmation means that the light curve of a supernova can be extracted from the GRB's light curve. Thus, the afterglow contribution and the host galaxy to the total flux observed from the source must be subtracted from the total GRB light curve. It is also necessary to take into account the absorption in both our [18] and the host galaxies, which will allow us to construct the correct light curve of the supernova associated with this gamma-ray burst.

In turn, spectroscopic confirmation of the GRB-SN relationship can be obtained as a result of spectroscopic observations in the region of the maximum of the supposed supernova. Typically, the spectra measured during the supernova phase are characterized by broad lines and the absence of hydrogen and helium lines, which is a distinctive feature of Type Ic supernovae.

To extract a supernova from the light curve of a gamma-ray burst, it is necessary to take into account all components from the afterglow power-law to absorption in the host galaxy. Also, without taking into account the absorption in the Galaxy and the contribution of the host galaxy to the overall light curve, it is impossible to obtain a correct SN-GRB light curve. To increase the statistics on this phenomenon, it is necessary to carry out this procedure of extraction for all GRBs, in which theoretically there could be a supernova. To date, of the total number of detected gamma-ray bursts, only 30 and 23 of them have received photometric and spectroscopic confirmation of a connection with a supernova, respectively (e.g. [19]). In this regard, each new confirmation of the GRB-SN connection still makes a significant contribution to the development of the science of this phenomenon.

2   Registration of GRBs and Their Observation

As it was mentioned earlier, the study of gamma-ray bursts experienced a noticeable rise after the launch of the Swift orbiting observatory [11]. This is due to the fact that Swift is the first autonomous fast-responding satellite for multi-wavelength transit astronomy. The Swift observations are carried out with three instruments: a wide-field gamma detector (15–150 keV), which is able to determine the position of the source with an accuracy of 1–4 arc minutes and start the automatic rotation of the spacecraft (spacecraft slew); a narrow-field X-ray telescope that can pinpoint the position of the source to within 5 arc seconds; and a UV/optical telescope that makes observations in the 170–600 nm wavelength range and provides localization accuracy of about 0.3 arc seconds. X-ray telescopes are needed because the localization of bursts in the hard X-ray and gamma ranges is not accurate enough for aiming optical telescopes. Accordingly, after receiving the coordinates of the source, they are transmitted to the Earth for organization of further observations at ground-based observatories.

A significant part of the results presented in this work was obtained as a result of organizing observations with small aperture telescopes, which are part of the IKI GRB Follow-up Network (IKI GRB-FuN; [20]) and with which our group actively interacts on the issue of observing transient events.

In addition to observations of the early stage of gamma-ray bursts, which includes both the active phase (if possible) and the afterglow, observations of the host galaxy are also organized when the source has already exhausted itself. This is done in order to correctly describe the overall light curve and estimate the contribution of the host galaxy to it. Observation data in some cases allow modeling the host galaxy in order to determine its physical parameters, such as type, mass, age, extinction, etc.

3   Astronomical Images Processing

As a result of the observations, we get astronomical images, which must then be prepared for use. A standard preliminary reduction is applied to them, which at the first step includes the rejection of defective images. Next, we apply correction to the remaining images with a bias matrix, a dark frame, and a flat field. Fringes, if any, are also removed. This processing is done in the IRAF[1] software package. The preliminary reduction was carried out by the *ccdproc* program of the *ccdred* package. After all the corrections have been applied to the astronomical frames, the individual images are summarized by *imcombine* in the *imgeom* package to improve the better signal-to-noise ratio. After that, the photometry of the source of interest is done using the *phot* program of the *apphot* package. As a result of this processing, we get the magnitude of our source, which we then put on the light curve to the points for other nights or in other filters. Carrying out the procedure for each series of observations gives the light curves that will be investigated further.

---

[1] IRAF (Image Reduction and Analysis Facility), an environment for image reduction and analysis, was developed and maintained by the National Optical Astronomy Observatory (NOAO, Tucson, USA) operated by the Association of Universities for Research in Astronomy (AURA) under cooperative agreement with the National Science Foundation of the USA, see iraf.noao.edu.

4   Sample

As a result of all the observations of a particular gamma-ray burst, one can try to extract a supernova sign from its light curve. Since a statistically reliable light curve can be plotted for most gamma-ray bursts, it is necessary to check most of them for the presence of a supernova sign in it.

As mentioned earlier, among the huge number of detected gamma-ray bursts, only a few dozen have received confirmation of connection with supernovae. Thus, the study of the GRB-SN relationship consists, in particular, in pulling out a grain of data useful for the development of the studied phenomenon from a large number of data/observations. The results of the performed procedure for constructing the supernova light curve are shown in the form of a matrix in Fig. 1.

Figure 1 shows the light curves of supernovae associated with gamma-ray bursts. These light curves were obtained from both our own data and the results of observations published in articles and are in the public domain. A separate light curve for each GRB was taken and processed, which will be described in the next section. As a result of this processing of each individual GRB light curve, the light curves of the supernova associated with GRBs were obtained, shown in Fig. 1.

This matrix is the result of the enormous work of many people who have contributed to all stages of work from the organization of observations, to their direct processing with obtaining results. It can also be seen from the matrix that, from 1998 to the present day, only a couple of dozen light curves of supernovae associated with gamma-ray bursts have been obtained. This fact still confirms the need to increase the statistics on GRB-SN cases and that each discovered SN in the light curve of the GRB still makes a significant contribution to the study of this phenomenon.

Another feature of this matrix is the variety of light curves and their uniqueness. No light curve is like the others. The matrix shows a wide range of possible GRB-associated supernova light curves that can be obtained. This diversity lies in the fact that the SN-GRB light curves have different growth and decay rates and, accordingly, are described by different exponential functions. Moreover, the time of the onset of the SN maximum and its amplitude for each SN are different: for some, the maximums are brighter than for others. The maximums of some of them may occur on fifth day, and for some on 25. And so far, no general regularities have been obtained that would be common to describe the variegated series of SN-GRB light curves.

The matrix (Fig. 1) shows the light curves of the following sequence of gamma-ray bursts: GRB 980425 [23], GRB 011121 [24], GRB 071112C [25], GRB 111209A [26], GRB 111228A [25], GRB 120422A [27], GRB 120714B [25], GRB 130702A [28], GRB 130831A [29], GRB 140606B [32], GRB 150818A, GRB 161219B [30], GRB 171010A [33], GRB 171205A [31], GRB 180728A, GRB 190114C [34], GRB 201015A [35].

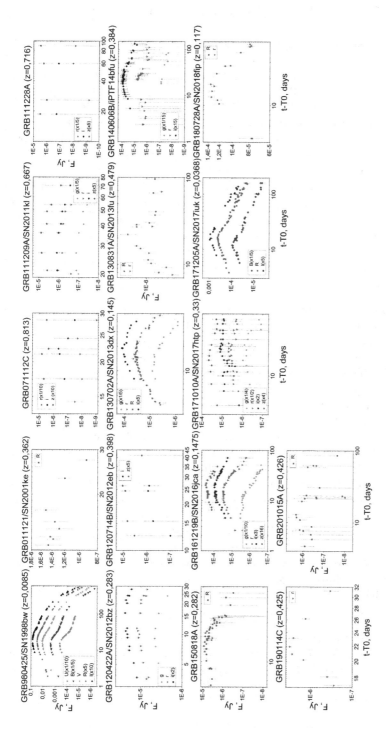

Fig. 1. A matrix of SN-GRB light curves. It is sorted according to the dates of the GRB registration. The redshift values for each SN-GRB are given in brackets. (Color figure online)

5 Light Curves of SNe Associated with GRBs

Obtaining a "long" GRB light curve depends on how long we can observe a GRB. Observations can last from a day to several months. It depends on the energy of the burst and on the distance to it. Also, observations of the unperturbed host galaxy of this particular GRB can be resumed after an even longer period of time, even after several years. This is necessary to build a full GRB light curve. If the burst has not exhausted itself on the very first day and can be observed for some time, then in the presence of a certain redshift of this burst, some conclusions can be drawn that may affect the planning of further observations. Today, it is known that about 70% of "long" GRBs with redshift less than about 0.5 have a supernova in itself. Therefore, if the detected gamma-ray burst had a redshift of less than about 0.5, then if other things being equal, it makes sense to organize its continuous observations in an attempt to detect a supernova in its light curve.

As a result of all the observations carried out, a light curve of a gamma-ray burst with a supernova sign is obtained. To obtain the supernova light curve, we have to do the following:

1. **Afterglow subtraction:** We approximate the afterglow by a power law in that part of the light curve where it is not perturbed by either the active phase, or the supernova, or host galaxy or other inhomogeneities. We then extrapolate the resulting power law to the entire range of the light curve and subtract it. Thus, we took into account the contribution of the afterglow to the overall light curve.
2. **Host galaxy subtraction:** Since the luminosity of the host galaxy changes negligibly over a time interval of several years, its contribution to the overall light curve can be considered constant. Observations of the unperturbed host galaxy will give us its flux, which must then be subtracted from the light curve remaining after subtracting the afterglow.
3. **Milky way extinction:** Taking into account the Galactic extinction is necessary to construct a correct supernova light curve, and it is determined for all directions from which the signal from the source is detected (see [18]).
4. **Host galaxy extinction:** An estimate of the extinction in the host galaxy can be obtained, in particular, by modeling the radiation of the host galaxy using the LePhare program code [21,22] developed to approximate the spectral energy distribution of galaxies and obtain their physical parameters. But it is not always possible to perform these manipulations, so the extinction in the host galaxy cannot always be taken into account.

As a result of the processing performed in the described way, a matrix was obtained, consisting of 17 light curves of supernovae associated with gamma-ray bursts, which is shown in Fig. 1. The light curves presented on the matrix can be phenomenologically divided into the following groups.

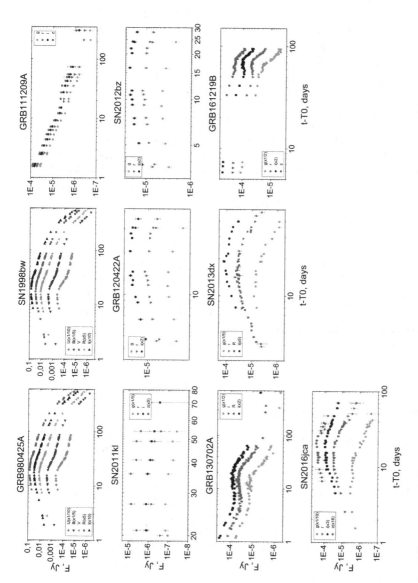

Fig. 2. Multicolor light curves of GRBs and its associated supernovae. The light curves of the GRB and the SN-GRB follow one after the other in rows.

5.1   Well Statistically Secure SN's Light Curve

This category includes those supernovae whose light curves were plotted in several photometric filters. That is, multicolor observations of both the afterglow, the supernova and the host galaxy stages were carried out. This made it possible to construct a statistically reliable supernova light curve with a significant number of points composing it. The light curves of such supernovae have maximum in several filters, which allows an explicit determination of such supernova parameters as the amplitude and position of the peak. In addition, physical modeling can be carried out, as a result of which such physical parameters of the supernova as the amount of radioactive ^{56}Ni synthesized during the explosion can be determined.

Figure 2 shows the light curves of GRBs (left column) and supernovae associated with them (right column). It can be seen that the dome of each supernova is well defined in at least 3 filters. The light curve of supernova SN 1998bw is the light curve of GRB 980425, since this burst, if it had an afterglow, is very weak, so it could be neglected, as well as the contribution of the host galaxy. The same is true for SN 2012bz (GRB 120422A).

5.2   Averagely Statistically Secure SN's Light Curve

It is not always possible to carry out continuous quasi-synchronous observations in several filters throughout the entire time of the activity of the GRB-SN. In particular, observations can be obtained in a pair of filters, which makes it possible to separate such SN-GRB light curves into a separate group. Often this situation is due to the fact that the time allotted for observing an object is limited on telescopes. Therefore, it is necessary to prioritize and conduct observations in a limited number of filters. An example of a light curve for a supernova of this type is GRB/SN 171010A (see Fig. 1).

5.3   Single-Filter SN's Light Curve

It is also possible that observations can be made in only one filter. In addition to the limitations associated with the distribution of time on the telescope, bad weather can also negatively affect the possibility of observing the GRB-SN. In this regard, multicolor observations are rare, and most often the observation of GRB-SN occurs in a single filter. Often these are the r and R filters of the AB and Vega photometric systems, respectively. And such cases are the prevailing number, so there is a division of similar GRB-SN into a separate group. This group of GRB-SN light curves includes, in particular, GRB 011121, GRB 130831A, GRB 150818A, GRB 180728A, GRB 190114C, GRB 201015A, which are shown in Fig. 1.

5.4   Subtle SN's Light Curve

Sometimes it may happen that many observations have been made both at the afterglow stage and at the supernova stage. And the oblong bulge associated with the supernova manifestation is visible on the overall light curve. But after subtracting all other components of the light curve (afterglow, host galaxy), it occurs that the supernova light curve cannot be identified as such. For example, there is no obvious dome or the errors are very large.

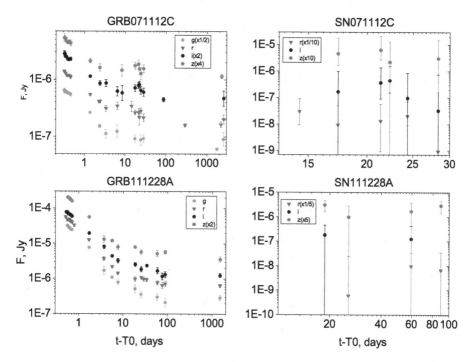

Fig. 3. Subtle light curves of SNe (right column) associated with GRBs (left column).

This situation can be explained by the fact that the source of the GRB-SN is located at a huge distance on the Earth, because of which the supernova manifests itself weakly, therefore it does not have a statistically guaranteed peak. These include, for example, GRB 071112C, the redshift of which is z = 0.812 (Fig. 3).

Also, a faint appearance of a supernova may be due to a bright afterglow, which makes a larger contribution to the total flux from the source than a supernova. As a result, the supernova is eclipsed by the afterglow. An example of such a situation is GRB 111228A (Fig. 3).

Another factor affecting the visibility of a supernova is its rather bright host galaxy. In this case, the supernova manifestation against the background of the power law decay of the afterglow will look insignificant. This is also the case for the aforementioned light curves, as shown in Fig. 3.

In addition to all this, the period and place of observations can affect the quality of the supernova light curve extraction. Consider GRB 181201A [19], for example. This source was observed for about 24 days and has provided afterglow stage in several filters. However, due to the fact that its culmination on about 20 days was spent on the daytime part of the day. In this regard, this source could only be observed with large aperture telescopes (e.g. 8.1-meter telescope Gemini-N), as well as those that look in the direction of the source through a smaller thickness of the atmosphere.

Thus, there are always some limiting factors that make it difficult to conduct comprehensive observations of supernovae associated with gamma-ray bursts. However, even from an incomplete dataset, it is often possible to extract a lot of useful data that can be used to advance research on this phenomenon.

5.5   Hybrid SN's Light Curve

This category includes light curves that do not fit the criteria of the previous types. For example, the light curve has many points, but their errors are large, which reduces the statistical significance of certain results. These include the light curve of the supernova associated with the GRB 140606B. The light curve of this supernova is well defined in the region of the dome (see Fig. 1), but the photometric points have large errors, which in some places even go to the negative values of flux. That does not allow attributing this light curve to any of the other types indicated above.

Thus, using the example of the presented light curves of supernovae associated with gamma-ray bursts, several classes were distinguished. This phenomenological division was based on the following factors:

- The number of photometric filters;
- The number of points at the stage of SN rise;
- The number of points in the area of the SN's dome;
- The number of points at the stage of SN decay;
- Contribution of the afterglow to the overall light curve of the GRB;
- Contribution of the host galaxy to the overall GRB's light curve;
- Distance to the SN-GRB source.

Depending on these parameters, the SN's light curve can be assigned to one of the groups presented above.

6   Description of Light Curves of SNe Associated with GRBs

After obtaining the supernova light curve, we want to know some of its parameters. These include the position of the supernova maximum and its maximum luminosity. For this, it is necessary to approximate the light curve with some analytical function. As such, the Bazin function [36], specially derived to describe the light curves of supernovae, can be used. It has a form that describes the exponential rise and fall of the supernova light curve. This function was obtained by a purely empirical method; therefore, it does not pretend to be a physical description of the light curve. This may be related to the fact that the approximation of the supernova light curve may not always be physically justified. Also, the Bazin function has five parameters, which, in the case of a small number of photometric points on the supernova light curve, makes it impossible to unambiguously fit it.

On the other hand, a simpler parabola function can be used [37]. As shown in the work [38], if it is necessary to approximate only the dome without describing the wings of the supernova light curve, then such analytical functions as a parabola, a polynomial of the fourth degree, and a lognormal distribution can be used. Each of them describes the light curve of a supernova as a whole differently, but they are all equally suitable for describing a supernova dome in the time range ±5 days from the supernova maximum. Thus, a parabola can be used to determine the position and amplitude of the supernova maximum. The advantage of using it is also that it has only three parameters, which makes it possible to unambiguously approximate the supernova light curve, in which there are only three points in one photometric filter.

An example of using a parabola as a function to approximate the dome of a supernova light curve is shown in Fig. 4.

Figure 4 shows the results of the parabola fit of the light curves of supernovae associated with gamma-ray bursts. For all light curves, one filter was taken with the largest number of photometric points. Although the parabola absolutely not applicable for supernova's wings description, it can be seen that it approximates the supernova dome quite well. Thus, we can obtain the parameters of the supernova (amplitude and position of the maximum) even in the case of a small number of photometric points at the supernova stage. For the light curves of supernovae associated with GRB 071112C and GRB 111228A, it was not possible to conveniently carry out the approximation procedure due to large errors.

7   Discussion

To date, a common practice for approximating GRB-associated supernova light curves is to fit the with statistically secure SN 1998bw and/or SN 2013dx light curve. This procedure consists in determining the stretch factor (s) and luminosity factor (k) to match the light curve of SN 1998bw or SN 2013dx (e.g. [39]).

Despite this, the question of unification of supernova light curves is still open. That is, there is no single criterion for describing all possible SN-GRB light curves. The classification of the SN-GRB cases given in this paper by their quality makes it possible to phenomenologically classify the supernova light curve to one of the described classes, which will help to understand how to work with it and what results can be expected.

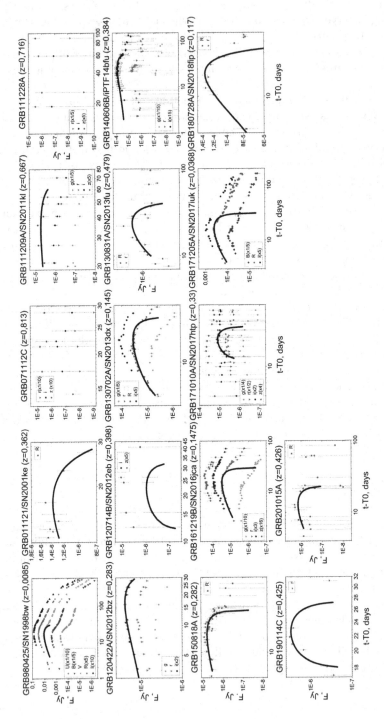

Fig. 4. Light curves of supernovae associated with gamma-ray bursts. The black solid line shows the best approximation of the light curves in one of the filters by a polynomial of the second degree (parabola).

For example, if the GRB is far away, then, as a rule, there is no need to expect a bright SN and, accordingly, it is necessary to be more careful both in planning observations and in data processing in order not to miss the SN-GRB.

8   Interdisciplinary Application

Mostly, the results of this work will be useful to those who are directly involved in the study of cosmic gamma-ray bursts with supernovae. But they also have a wider range of potential applicability and benefits to other areas of science. Firstly, this is directly related to technical development. Today we can observe SN-GRB, for the most part no further than a redshift of 0.5–0.9. Accordingly, an increase in statistics on SN-GRB requires, on the one hand, long and quasi-continuous monitoring of the sky, and, on the other hand, the technical development of telescopes in order to be able to conduct deeper observations. In response to this request, more technically advanced telescopes will be developed. It will help register more distant supernovae, which will be useful in increasing statistics on SN-GRB.

Secondly, deeper observations will allow looking at earlier stages of the Universe, which is important for the development of cosmology. Such progress in the observational ability of telescopes will make it possible, for example, to refine the Hubble constant or to determine the constraints on the cosmological constant. The latter, in turn, is inextricably linked to the expansion of the universe and dark energy which makes up to 68% of our Universe. Therefore, this direction of development is important for understanding the world in which we live. With no doubt, the development of astrophysics will directly affect physics, since the study of high-energy objects, such as gamma-ray bursts, supernovae, neutron stars, will help to understand those processes that take place in them that cannot be reproduced in terrestrial conditions.

Thirdly, deeper observations will expand the understanding of what place the Earth and humanity occupy in this world, which will give even more ground for thought to philosophers. And new methods of researching a phenomenon will allow to continue discussions about the method of scientific cognition, which is now firmly established, but may possibly lead to a paradigm shift, which will give a new leap in scientific knowledge.

Thus, we can conclude that any laboratory experiments are about accuracy, while astrophysical experiments are about a range, and this possibility, of course, cannot be neglected.

9   Conclusion

If summarise everything that was mentioned above it is important to say that out of the huge set of detected gamma-ray bursts over more than 50 years, only a few dozen of them received photometric or spectroscopic confirmation of a connection with a supernova. The light curve of each supernova, in turn, was obtained as a result of extensive processing of a colossal number of observations. This requires continuous observations of both the afterglow and the supernova phases. Observations of the unperturbed host galaxy are also important for correct deconstruction of the SN-GRB light curve. Statistics on this phenomenon is still at the initial stages, therefore, it is necessary to introduce their classification already at this stage for its further development. This will certainly happen for the reason that the registration of supernovae associated with gamma-ray bursts occurs with a frequency of 1–3 per year. However, with the development of the observing apparatus, which will allow registering fainter SN or SN from more distant sources than nowadays, this frequency will only increase and perhaps someday it will reach 10 SN per year or even more.

The matrix presented in this paper shows that each SN-GRB light curve is unique and not similar to the others. This can be observed by plotting the distribution of SN-GRB over the peak width in days (Fig. 5). The peak width was calculated at 87.3% of the supernova maximum flux. This level was taken from the parabola fit of the one of a few statistically reliable light curve of SN 2013dx associated with GRB 130702A (see [19]). At this level, from the SN maximum, the parabola approximation began to diverge from the experimental points. This distribution of the peak width is shown in Fig. 5. The mean of the distribution is ~12.1 and RMS is ~13.1 days.

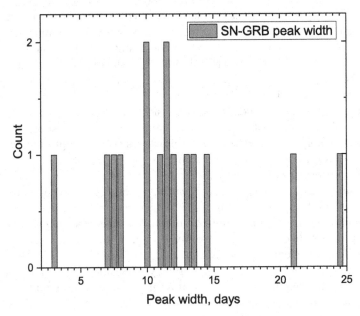

Fig. 5. Distribution of SN-GRB light curves presented in this paper over the width of the dome.

A comparison between supernovae associated with GRBs and SNe Ic can be made. Figure 6 shows the distribution of SN-GRB and SN Ic in absolute magnitude at the supernova maximum. The SN-GRB distribution for this parameter was constructed based on the supernova light curves presented in this paper. The SN Ic distribution was taken from the paper of Richardson ([40]), and for its construction the Asiago Supernova Catalogue (ASC, [41]) mostly was used.

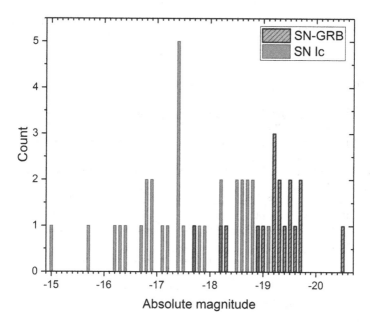

Fig. 6. Comparison of the distributions of SN-GRB and SN Ic in absolute magnitude at the maximum of the light curve.

From this distribution (Fig. 6), it can be seen that supernovae associated with gamma-ray bursts are predominantly brighter than type Ic supernovae. Moreover, according to the distribution of SN-GRB in absolute values, it can be seen that they lie in the range from -17.6 to -20.6 and only in some cases have approximately the same magnitudes.

For these two datasets, the Two-sample Kolmogorov-Smirnov test (KS-test) was carried out. As a result, it does not reject the null hypothesis that the samples SN-GRB and SN Ic belong to the same general distribution (p-value = 0.3549). Thus, it is necessary to increase the number of discovered SN-GRB in order to be able to statistically divide them into a separate general distribution from the SN Ic.

A distinctive feature of the SN-GRB light curves is also the fact that, unlike other types of supernovae, they do not have a plateau phase.

The matrix will develop as the existing data are processed and new data on this phenomenon are obtained. This will ultimately make it possible to either find some common feature for all supernovae associated with gamma-ray bursts, or it will confirm the property of the uniqueness of each light curve. The second would mean that any use of templates for describing SN-GRB light curves can be performed only in some approximation, and also that each light curve needs an individual approach and they are all worthy of a separate detailed consideration. The main conclusion of SN light curve discussed that the number of SNs associated with GRB is still not complete to secure statistical investigations and determine possible difference of SN Ic and SN associated with GRB.

Acknowledgements. Authors acknowledge the support of the RSCF grant 18-12-00378.

References

1. Klebesadel, R.W., et al.: Observations of gamma-ray bursts of cosmic origin. Astrophys. J. **182**, L85 (1973)
2. Mazets, E.P., et al.: Catalog of cosmic gamma-ray bursts from the KONUS experiment data. Astrophys. Space Sci. **80**(1), 3–83 (1981)
3. Paciesas, W.S., et al.: The fourth BATSE gamma-ray burst catalog (revised). Astrophys. J. Suppl. Ser. **122**(2), 465–495 (1999)
4. Meegan, E.C., et al.: Spatial distribution of γ-ray bursts observed by BATSE. Nature **355**(6356), 143–145 (1992)
5. Paczynski, B.: How far away are gamma-ray bursters? Publ. Astron. Soc. Pac. **107**, 1167 (1995)
6. Paczynski, B., Rhoads, J.E.: Radio transients from gamma-ray bursters. Astrophys. J. Lett. **418**, L5 (1993)
7. Costa, E., et al.: Discovery of an X-ray afterglow associated with the γ-ray burst of 28 February 1997. Nature **387**(6635), 783–785 (1997)
8. van Paradijs, J., et al.: Transient optical emission from the error box of the γ-ray burst of 28 February 1997. Nature **386**(6626), 686–689 (1997)
9. Frail, D.A., et al.: The radio afterglow from the γ-ray burst of 8 May 1997. Nature **389**(6648), 261–263 (1997)
10. Frontera, F., et al.: The gamma-ray burst catalog obtained with the gamma-ray burst monitor aboard BeppoSAX. Astrophys. J. Suppl. **180**(1), 192–223 (2009)
11. Gehrels, N., et al.: The swift gamma-ray burst mission. Astrophys. J. **611**(2), 1005–1020 (2004)
12. Swift GRB table. https://swift.gsfc.nasa.gov/archive/grb_table/fullview/. Accessed 26 Feb 2018
13. von Kienlin, A., et al.: The fourth fermi-GBM gamma-ray burst catalog: a decade of data. Astrophys. J. **893**(1), 14, Article ID 46 (2020)
14. Kouveliotou, C., et al.: Identification of two classes of gamma-ray bursts. Astrophys. J. Lett. **413**, L101 (1993)
15. Nakar, E.: Short-hard gamma-ray bursts. Phys. Rep. **442**(1–6), 166–236 (2007)
16. Woosley, S.E., Bloom, J.S.: The supernova gamma-ray burst connection. Ann. Rev. Astron. Astrophys. **44**(1), 507–556 (2006)
17. Mazaeva, E., et al.: Inhomogeneities in the light curves of gamma-ray bursts afterglow. Int. J. Mod. Phys. D **27**(10), Article ID 1844012 (2018)
18. Schlafly, E.F., Finkbeiner, D.P.: Measuring reddening with Sloan digital sky survey stellar spectra and recalibrating SFD. Astrophys. J. **737**(2), 13, Article ID 103 (2011)
19. Belkin, S.O., et al.: Multiwavelength observations of GRB 181201A and detection of its associated supernova. Astron. Lett. **46**(12), 783–811 (2020). https://doi.org/10.1134/S1063773720120014
20. Volnova, A., et al.: Proceedings of the 22nd International Conference DAMDID/RCDL-2020, Voronezh, Russia, 13–16 October 2020, Ed. by Kozaderov, O.A., Zakharov, V.N. CCIS (2020, in press)
21. Arnouts, S., et al.: Measuring and modelling the redshift evolution of clustering: the Hubble deep field north. Mon. Not. **310**(2), 540–556 (1999)

22. Ilbert, O., et al.: Accurate photometric redshifts for the CFHT legacy survey calibrated using the VIMOS VLT deep survey. Astron. Astrophys. **457**(3), 841–856 (2006)

23. Clocchiatti, A., et al.: The ultimate light curve of SN 1998bw/GRB 980425. Astron. J. **141**(5), 6, Article ID 163 (2011)

24. Garnavich, P.M., et al.: Discovery of the low-redshift optical afterglow of GRB 011121 and its progenitor supernova SN 2001ke. Astrophys. J. **582**(2), 924–932 (2003)

25. Klose, S., et al.: Four GRB supernovae at redshifts between 0.4 and 0.8. The bursts GRB 071112C, 111228A, 120714B, and 130831A. Astron. Astrophys. **622**, 28, Article ID A138 (2019)

26. Kann, D.A., et al.: Highly luminous supernovae associated with gamma-ray bursts. I. GRB 111209A/SN 2011kl in the context of stripped-envelope and superluminous supernovae. Astron. Astrophys. **624**, 19, Article ID A143 (2019)

27. Melandri, A., et al.: The optical SN 2012bz associated with the long GRB 120422A. Astron. Astrophys. **547**, 7, Article ID A82 (2012)

28. Volnova, A.A., et al.: Multicolour modelling of SN 2013dx associated with GRB 130702A. Mon. Not. R. Astron. Soc. **467**(3), 3500–3512A (2017)

29. Cano, Z., et al.: A trio of gamma-ray burst supernovae: GRB 120729A, GRB 130215A/SN 2013ez, and GRB 130831A/SN 2013fu. Astron. Astrophys. **568**, 16, Article ID A19 (2014)

30. Cano, Z., et al.: GRB 161219B/SN 2016jca: a low-redshift gamma-ray burst supernova powered by radioactive heating. Astron. Astrophys. **605**, 21, Article ID A107 (2017)

31. Volnova, A.A., et al.: In preparation (2021)

32. Cano, Z., et al.: GRB 140606B/iPTF14bfu: detection of shock-breakout emission from a cosmological γ-ray burst? Mon. Not. R. Astron. Soc. **452**(2), 1535–1552 (2015)

33. Melandri, A., et al.: GRB 171010A/SN 2017htp: a GRB-SN at z = 0.33. Mon. Not. R. Astron. Soc. **490**(4), 5366–5374 (2019)

34. Pozanenko, A.S., et al.: In preparation (2021)

35. Belkin, S.O., et al.: In preparation (2022)

36. Bazin, G., et al.: Photometric selection of type Ia supernovae in the supernova legacy survey. Astron. Astrophys. **534**, 21, Article ID A43 (2011)

37. Bianco, F.B., et al.: Multi-color optical and near-infrared light curves of 64 stripped-envelope core-collapse supernovae. Astrophys. J. Suppl. **213**(2), 21, Article ID 19 (2014)

38. Belkin, S.O., et al.: Proceedings of the XXIst International Conference DAMDID/RCDL-2019, Kazan, Russia, 15–18 October 2019, Ed. by Elizarov, A., Novikov, B., Stupnikov, S., vol. 2523, pp. 244–254 (2019)

39. Cano, Z., et al.: XRF 100316D/SN 2010bh and the nature of gamma-ray burst supernovae. Astrophys. J. **740**(1), 17, Article ID 41 (2011)

40. Richardson, D., et al.: Absolute-magnitude distributions of supernovae. Astrophys. J. **147**(5), 17, Article ID 118 (2014)

41. Barbon, R., et al.: The Asiago supernova catalogue. Astron. Astrophys. Suppl. Ser. **81**, 421–443 (1989)

Application of Machine Learning Methods for Cross-Matching Astronomical Catalogues

Aleksandra Kulishova[1](✉) [ID] and Dmitry Briukhov[2] [ID]

[1] Lomonosov Moscow State University, Moscow, Russia
sasha_kulishova@mail.ru
[2] Federal Research Center "Computer Science and Control" of Russian Academy of Sciences,
Moscow, Russia

Abstract. The paper presents an approach for the application of machine learning methods for cross-matching astronomical catalogues. Related works on the cross-matching are analyzed and machine learning methods applied are briefly discussed. The approach is applied for cross-matching of three catalogues: Gaia, SDSS and ALLWISE. Experimental results of application of several machine learning methods for cross-matching these catalogues are presented. Recommendations for the application of the approach in astronomical information systems are proposed.

Keywords: Machine learning · Cross-matching · Cross-identification · Astronomy

1 Introduction

Since ancient times the main method for obtaining the data on celestial objects in astronomy was observation of the night sky. Most astronomical problems are solved using the analysis of the observational data. The night sky is surveying at many astronomical observatories around the world. In addition, special space telescopes, such as Hubble [1], Gaia [2], and IRAS [3], are being built to improve the accuracy of observations and reduce the effects of the Earth's atmosphere. As a result, large arrays of astronomical images were obtained. Each image within these arrays contains data about several thousand astronomical objects. The data is usually extracted from images and collected in astronomical catalogues in a systematic way. The catalogues can contain data about objects in a certain area of the sky or the entire sky. The set of characteristics contained in a catalogue depends on the observational facilities and research goals. Catalogues can contain data received at different times (epochs) and in different passbands. For instance, the Hubble telescope operates the optical passband, while the IRAS telescope operates the infrared passband. The telescopes have different sensitiveness following with different astrometric and photometric errors.

Consequently, the same astronomical object in different catalogues may have different characteristics. For instance, an object with a high proper motion included in two different catalogues can be observed at different epochs and in different passbands.

© Springer Nature Switzerland AG 2022
A. Pozanenko et al. (Eds.): DAMDID/RCDL 2021, CCIS 1620, pp. 92–103, 2022.
https://doi.org/10.1007/978-3-031-12285-9_6

Such differences complicate matching the data on the same astronomical objects from different catalogues. This problem arises in many application use cases, such as:

- *Integration of several catalogues* for usage in an astronomical information system.
- *Astronomical data analysis* problems often require data from several catalogues due to their difference in characteristics and coverage of sky regions. For example, in [4] the authors apply a comparison of the WISE [5] and the SuperCOSMOS [6] catalogues to solve the galaxies identification problem.
- *Identification of the differences between catalogues* is used to search for objects with high proper motion or new objects that were not previously identified in other catalogues.

The problem of matching catalogues to find the same objects is called the *cross-matching astronomical catalogues*. Several different approaches for solving this problem exist. As the popularity of machine learning methods grows, the researchers begin to apply them to different astronomical problems including cross-matching catalogues. The aim of this paper is to investigate the application of various machine learning methods for cross-matching objects in astronomical catalogues and to propose a generalized approach to solve the problem.

The paper is focused on methods for matching two catalogues, since the matching of several catalogues can be reduced to several sequential matchings of two catalogues. Similar approach is applied, for instance, in [7, 8]: one of the catalogues is designated as the "central" or the "main" one, and it is matched with the all other catalogues.

Cross-matching of two catalogues A and B starts with the creation a set of potential pairs (X, Y) of the matched objects where X is an object from the catalogue A and Y is an object from the catalogue B. To do this, the entire sky is divided into regions of a given size. Objects from a region are extracted from both catalogues and combined into pairs. Each pair of objects contains a list of attributes (characteristics) retrieved from the catalogues. The minimal set of attributes includes coordinates of objects and the flux values for different passbands. After that, the cross-matching problem can be considered as a binary classification problem in machine learning. For each pair, it is required to tell whether its objects refer to the same astronomical object or not, in other words whether the objects of a pair are matched or not.

The paper is organized as follows. Section 2 provides an overview of recent existing cross-matching approaches. Section 3 describes the proposed approach for solving the problem of cross-matching of astronomical catalogues. Section 4 presents the comparative results of different machine learning methods application for cross-matching objects in Gaia, ALLWISE [9], and SDSS [10] catalogues.

2 Related Work

To develop a generalized approach for cross-matching astronomical catalogues proposed in the next section we reviewed the related existing approaches for solving the problem first. The possible complications are identified; advantages and disadvantages of the related approaches are considered.

In [11] the authors describe the problem of matching two catalogues (HOPCAT and SuperCOSMOS) using machine learning algorithms. The paper is focused on choosing the machine learning model and the features (characteristics) for it. To select features required for machine learning, the authors propose to search for correlations between the characteristics of objects from different catalogues. Simple preprocessing of the data is performed. The most appropriate machine learning algorithm to be applied for cross-matching is defined using the cross-validation procedure. The data is divided into n equal parts, then the algorithms are trained on $(n - 1)$ parts of them and are tested on the rest part. The procedure is repeated n times with different test subsets. Results are averaged for each algorithm. Two best algorithms for the matched catalogues were identified: the SVM with 99.12% correct answers, and the multilayer perceptron with 95.2–96% correct answers.

In [7], the authors describe an algorithm adopted for cross-matching the Gaia Data Release 2 (DR2) catalogue with large dense and sparse catalogues. Changes and improvements of the algorithm comparing with one adopted for Gaia DR1 are described in details [8]. The results of the work became a part of the Gaia DR2 catalogue. The authors match Gaia with each external catalogue independently. So the Gaia catalogue is considered as the "central" one. Matches across multiple external catalogues are found as combinations of their matches with Gaia. For large dense surveys, Gaia is the leading catalogue, while for sparse catalogues, the external catalogue leads in cross-match.

For a given object in the leading catalogue the search for a counterpart objects in the matched catalogue (called the *second* one) includes two stages. First, "good neighbors" are found that are nearby objects in the second catalogue whose position is compatible within position errors with the target. Second, the "best neighbor" is found from these "good neighbors" (i.e. the most probable counterpart). The authors apply the same cross-matching algorithm as in [8]: a plane-sweep technique. The search for good neighbors is performed using two filters. The first filter calculates the initial search radius and is used only to select suitable candidates for good neighbors. The second filter is based on the Mahalanobis normalized distance and is used to select the good neighbors from candidates. The choice of the best neighbor from good neighbors is based on a specific figure of merit, which depends on the angular distance, position errors, epoch difference, and the local surface density of the second catalogue.

The MatchEx method for cross-matching catalogues is described in [12]. It is a rule-based method and it was deployed for the NED extragalactic database. To select the most appropriate candidate the MatchEx uses object positions, position uncertainties, types, names, background object density. This method uses conditional logic to determine which NED objects are acceptable matches. The match criteria include thresholds on separation, normalized separation, Poisson probability, inclusions or exclusions of types and names. The authors matched the GASC catalogue including nearly 40 million objects, with the NED catalogue including 180 million objects. The authors estimate the match completeness of 97.6%, and the accuracy of 99%.

In [13], the authors suggest an approach for cross-matching of multiple catalogues based on the integer linear programming. The approach extends previous work [14] following the idea to union all objects from different catalogues and to split this set into disjoint subsets of objects. In [14], likelihood optimization was compared with discrete minimization based on the Hungarian algorithm [15]. In [13] the integer linear programming approach was tested for the cross-matching two simulated catalogues of point sources. It worked in the same way as the previous method using the Hungarian algorithm, since both algorithms optimize the same problem. Matching the three catalogues is much more computationally complicated and the Hungarian algorithm is not applicable. Comparing to the method from [14], the integer linear programming approach can be extended on several catalogues because it does not require any assumptions on the number of catalogues.

The paper [16] describes a cross-matching solution underlying the computing infrastructure of the VASCO project, whose goal is to search for astronomical anomalies in historical all-sky surveys. An anomaly here means an inconsistency in sky surveys. The paper matches the USNO-B1 and the PanSTARRS1 catalogues. The idea of the algorithm is as follows. The selected area on the sky is divided into small windows of a given size. For each window, a table with pairs of objects matched with a specified precision is created, and another table with mismatching objects is created. For each mismatching object in the leading catalogue a mismatched pair is stored containing the closest object in the matched catalogue. The proposed cross-matching method applies the Euclidean distance. A match is a pair with the distance value that does not exceed the specified threshold. The author discovered 426,975 potential mismatches in the initial full-sky survey. Thus it was has shown that even with such a simple method decent results can be achieved. However, the issues of optimal memory usage, optimization of calculations to increase the performance, and preprocessing of the initial data still remain urgent.

In [17] the authors present an algorithm for efficient cross-matching, which utilizes only the coordinates of objects. This method calculates the increment of the number of stars in nested circular zones of increasing radii. The sum of two independent functions is used for this calculation: the first one describes the uniform density distribution of stars over the sky; the second one is the density distribution function of angular distances for the nearest neighbors. The method was used to match the PMA and the PPMXL catalogues, and others. Additionally, the authors analyze the cross-matching methods used in modern astronomic information systems. For instance, in the CDS cross-matching service, calculations take into account only objects whose positions are at an angular distance less than a specified radius. The main problem of this method is the incorrect matching of objects in areas with high density, when the search window includes more than one object. It is also noted that existing algorithms are performance-intensive. To overcome these issues, the authors sort the input data and apply the multithreading of the CPUs to speed up the calculations. This paper illustrates that existing position-based algorithms can be not accurate enough and should be optimized w.r.t. used memory and performance.

3   A Machine Learning Approach for Cross-Matching Astronomical Catalogues

The following machine learning methods were chosen to be applied for cross-matching astronomical catalogues on the basis of related work analysis: decision trees algorithm, support vector machine (SVM), multilayer perceptron, gradient boosting methods including a stochastic gradient boosting ensemble (LightGBM) and gradient boosting on trees (XGBoost)[1]. To compare the quality of the algorithms on cross-matching problem, the ε-neighborhood method was implemented to serve as the baseline. This method applies the Euclidean distance to match the objects on the basis of their coordinates.

3.1 The Machine Learning Approach Overview

The proposed machine learning approach for cross-matching two astronomical catalogues (called *the first* and *the second* respectively) consists of the following stages:

1. *Feature selection.* The required characteristics of object are extracted from each catalogue. These characteristics include the object coordinates and the flux in different passbands used in specific catalogues.
2. *Input data preparation.* The values of the selected characteristics are converted to the required types. The set of possible matching pairs is created. The train and the test sets of matching pairs are created. The noisy data are generated if required.
3. *Applying selected machine learning methods.*

   3.1. Selection of the parameters for applied machine learning methods.
   3.2. Machine learning models training.
   3.3. Application the positional comparison (ε-neighborhood) method for the labeled data.
   3.4. Evaluation of quality metrics, result comparison, selecting the best model.
   3.5. Application of the best model to unlabeled data.

The required set of features is extracted from the characteristics coming from both catalogues. These data are divided into windows of a given size on the basis of coordinates. After that each object from a window of the first catalogue is paired with each object of the second catalogue from the same window. This set is passed to the input of the already trained machine learning model, which solves the problem of binary classification, i.e. deciding whether the paired objects represent the same physical object or different physical objects.

For training machine learning models and for selecting the best model, the train and test sets are created on the basis of some pre-labeled set of pairs. The test set is used to compare the results of selected machine learning methods applying a given set of quality metrics. As the baseline the ε-neighborhood method is used during comparison.

For the matching more than two catalogues, it is proposed to select the leading catalogue and match all the rest catalogues with it in the given areas in the same way as in [7].

---

[1] Software implementation of the algorithms was taken from *sklearn* [18], *lightgbm*, and *xgboost* Python libraries.

3.2 Data Sources Selection

To illustrate the approach three catalogues that are Gaia [2], SDSS [10], and ALLWISE [9] were chosen as far as the data on cross-matching objects in these catalogues is open and available.

The Gaia is a space telescope of the European Space Agency, which mission is to provide a three-dimensional map of the Milky Way Galaxy, reveal its composition and evolution [19]. Before the publication of new versions of the Gaia catalogue, the authors cross-match it with well-known astronomical catalogues, such as SDSS, Pan-STARRS1, 2MASS, and others. This data is available on the Gaia website [2]. The Gaia currently has three officially published data releases. In this work, data on the position of objects (coordinates) and the flux values in three passbands from the Gaia DR2 are used.

The mission of SDSS (Sloan Digital Sky Survey) is to create the most detailed and high-quality three-dimensional maps of the Universe [10]. The ninth version of the data is the first publication of spectra from the SDSS-III baryon spectroscopic survey including more than 800,000 spectra over an area of 3,300 square degrees on the sky. In this work, the coordinates of the objects and the flux values in five filters (ultraviolet, green, red, near infrared, infrared) from the SDSS DR9 were used.

The ALLWISE project is based on the results of the successful WISE mission by combining two complete sky coverage epochs [9]. The ALLWISE catalogue is the most comprehensive mid-infrared view of the entire sky currently available. The catalogue contains accurate positions, apparent motion measurements, four-band fluxes, and flux variability statistics for over 747 million objects. In this work, the data on the position of objects and four-band fluxes are used.

3.3 Data Preparation and Data Quality Issues

Depending on the quality of data in matched catalogues several cases are considered:

1. The data in matched catalogues have been cleaned, the objects were observed at the same epoch approximately, there are no *nearby* different objects (whose positions are compatible within position errors).
2. The catalogues include data on objects observed in approximately the same epoch as well as data on objects observed in different epochs. There are some nearby different objects.
3. One of the catalogues is noisy or includes a large number of objects with high proper motion observed in a different epoch than in the other catalogue.

The first case is covered by matching the Gaia DR2 and the SDSS DR9 catalogues. Train and test sets were created on the basis of the *sdssdr9_best_neighbour*[2] from the Gaia DR2 archive [2]. The table contains about 113 million correlations, even though SDSS DR9 contains about 469 million objects, and Gaia DR2 contains about 1.6 billion objects. The table contains almost no examples of nearby different objects, and coordinates of the matched objects in these catalogues are almost the same (pairs of such

---

[2] https://gea.esac.esa.int/archive/documentation/GDR2/Gaia_archive/chap_datamodel/sec_dm_crossmatches/ssec_dm_sdssdr9_best_neighbour.html.

objects are called as "good" below). The training and test sets were created randomly, without any additional conditions.

For the second and third cases, the Gaia DR2 and the ALLWISE catalogues were chosen for the cross-matching. The train and test data sets were created on the basis of the *allwise_best_neighbour*[3] table from the Gaia DR2 archive [2] that contains 300 million objects matches for these catalogues. The known matches and mismatches of objects from the mentioned catalogues were collected from different regions of the sky. Two types of mismatches were considered: (1) mismatching objects that are located at sufficient distance from each other and (2) nearby mismatching objects. Pairs of such objects are called as "bad" below. Several train and test datasets were created containing from 1000 to 10000 pairs of objects with a different ratio of "bad" data. Preliminary experiments showed that on the datasets with a fairly low percentage of "bad" data, the selected methods showed results that are very similar to the results for the first case.

As far as the Gaia DR2 and the ALLWISE catalogues include a small number of nearby different objects and objects observed at the different epochs, it was decided to emulate the third case by adding some noise to the coordinates of one of the ALLWISE catalogue.

3.4 Train Set Creation Issues

One of the important sub-problems associated with the astronomical catalogues matching based on machine learning methods is to create the train set correctly. Within this work the following features and conditions for train set construct were identified:

- Train set may be small, about 500–1500 objects.
- Objects should be distributed over the entire coordinate range in catalogs.
- Train set should contain both nearby different objects and same objects that are located at sufficient distance from each other in different catalogues.
- The number of examples of mismatching objects should be more than the number of matching objects (which is close to real data).

Depending on the size and the structure of the data, the number of both nearby different objects and same distant objects should be at least 10–15% of the total number of objects. The experiments show that most machine learning methods have the same quality as the ε-neighborhood method otherwise, they match all objects located closely enough. The upper limit of such objects in a train set depends on their rate in a catalogue. In the selected catalogues the rate is not more than 20–25%.

The selected datasets had a drawback i.e. the number of "bad" pairs is much less than "good" ones, so the rate of such pairs in a train set should be increased. It can be achieved in two ways:

- *Add more "bad" data.* One should not use an entire set of "bad" data in the catalogue or face the problem of overfitting machine learning algorithms on these data otherwise.

---

[3] https://gea.esac.esa.int/archive/documentation/GDR2/Gaia_archive/chap_datamodel/sec_dm_crossmatches/ssec_dm_allwise_best_neighbour.html.

- *Reduce the number of "good" examples.* In this case, the data volume may not be enough to train the models up to the decent quality.

A combination of both ways was applied within this work to achieve better values of quality metrics.

4   Experiments and Results

This section describes the experiments on application of machine learning methods and the ε-neighborhood algorithm for cross-matching astronomical catalogues. The experiments were carried out for three datasets: (1) a dataset containing cleaned data without nearby different objects, (2) a dataset containing nearby different objects, and (3) a dataset containing noisy data. For each case, several experiments with different train and test sets were performed. Table 1 shows the number of experiments and the rate of "good" and "bad" objects in experiments for each case. It also shows the interval of noise for case 3. The train and test sets were created using the approach described in Sect. 3.

Table 1. The experiments parameters for different datasets.

Datasets	Number of experiments	Number of "bad" data (%)	Number of "good" data (%)	Interval of noise (arcsec)
Cleaned data	5	0.001%	99.999%	0
Nearby different objects	25	20–27%	73–80%	0
Noisy data	20	15–25%	75–85%	$[-2.5, 2.5]$

Values of quality metrics obtained for the selected machine learning algorithms are compared with the respective values obtained for the ε-neighborhood method. Three mentioned cases are considered one by one below.

The Objects in Catalogues were Observed at the Same Epoch Approximately, There are Almost no Nearby Different Objects. Several train and test datasets with different numbers of pairs of matching objects were selected from matching table of the Gaia DR2 [2] and the SDSS DR9 [10] catalogs randomly. Most of the machine learning models showed almost 100% matching quality, but the ε-neighborhood method also showed the 100% quality. However, the SVM and the K-nearest neighbors models showed the accuracy about 50–70% for different datasets.

The Catalogues Include Nearby Different Objects Closely or Same Objects Observed in Different Epochs. Several train and test datasets with different percentage of "bad" data were selected from the matching table of the Gaia DR2 [2] and ALLWISE [9] catalogues. Every dataset contains 5000 pairs of objects with 25–27% of them that are "bad" ones.

Quality metrics for the ε-neighborhood method depends on the percentage of the "bad" data in the train set. The results are shown in the Table 2.

Table 2. Quality metrics for the ε-neighborhood method applied to the dataset with nearby different objects or same objects observed in different epochs.

Quality metric	Metric value (%)	
	$\varepsilon = 0.01$ arcsec	$\varepsilon = 0.1$ arcsec
Accuracy	85–86.5	86–87.5
Precision	98–100	81–85
Recall	75–77.5	98–99.5
F1-score	83–86.5	89–91

Table 3 contains the results of application of machine learning methods for this dataset. The decision trees and gradient boosting methods showed better results than the ε-neighborhood method. The rest of the models showed not so good quality.

Table 3. The results of applying machine learning methods for the dataset with nearby different objects or same objects observed in different epochs.

Quality metric	Method	Metric value (%)
Accuracy	**DecisionTreeClassifier, XGBClassifier**	**88–97**
	LGBMClassifier	88–93
	KNeighborsClassifier	68–71
	SVM, MLPClassifier	75–80
Precision	**DecisionTreeClassifier, XGBClassifier**	**98–100**
	LGBMClassifier	**98–100**
	KNeighborsClassifier	65–67
	SVM, MLPClassifier	70–76
Recall	**DecisionTreeClassifier, XGBClassifier**	**89–96**
	LGBMClassifier	80–90
	KNeighborsClassifier	89–93
	SVM, MLPClassifier	91–96.5

One of the Catalogues is Noisy. As in the previous case, several train and test datasets with different percentage of "bad" data and different noise level were selected from the matching table of the Gaia DR2 [2] and ALLWISE [10] catalogues. The noise from the specified interval was added to the coordinates of objects from the ALLWISE catalogue: for the training set, it was randomly taken from the interval $[-1, 1]$ arcseconds. For the test set, with the size and percentage of "bad" data similar to the test set described above, the error was taken from the interval $[-2.5, 2.5]$ arcseconds.

As expected, the ε-neighborhood method works worse than on the previous dataset because of errors in coordinates. For small ε the method didn't match the same objects which coordinates are quite different. For the large ε it incorrectly matches different objects that are closely located. The results of application of the ε-neighborhood method to the noisy dataset are presented in Table 4.

Table 4. Evaluation of the ε-neighborhood method for the dataset with noise data.

Quality metric	Metric value (%)		
	$\varepsilon = 0.1$ arcsec	$\varepsilon = 1$ arcsec	$\varepsilon = 2$ arcsec
Accuracy	42–44.5	44–50	60–63
Precision	49–50	60–72	73–74
Recall	1–2	11–13	48–50
F1-score	1–2	20–24	57–60

Table 5. Results of applying machine learning methods for the dataset with noise data.

Quality metric	Method	Metric value (%)
Accuracy	**DecisionTreeClassifier, XGBClassifier**	**80–87**
	LGBMClassifier	81–84
	KNeighborsClassifier, MLPClassifier	69–74
	SVM	61–64
Precision	DecisionTreeClassifier, XGBClassifier	**79/86–92**
	LGBMClassifier	**83–92.5**
	KNeighborsClassifier, MLPClassifier	64–70
	SVM	65–67
Recall	DecisionTreeClassifier, XGBClassifier	**74–90**
	LGBMClassifier	74–83.5
	KNeighborsClassifier, MLPClassifier	87–92
	SVM	68–69

As it shown in Table 5 the decision tree and gradient boosting models worked well for the noisy data, but worse than for the previous dataset. The algorithms such as K-nearest neighbors, multilayer perceptron, and support vector machines are less sensitive to the noise. For example, for the K-nearest neighbors, the accuracy metrics are almost the same as for the previous dataset.

Experimental results of cross-matching astronomical catalogs can be summarized as follows. Both the conventional ε-neighborhood method and machine learning methods have their advantages and disadvantages.

Advantage of the ε-neighborhood method that it requires only data on the coordinates of objects and does not require data preparation and model training. It works well for cleaned catalogs that do not contain nearby different objects. However, its application meets several problems. Depending on the volume of data, calculations may require a long time and high resource costs. The value of ε should be carefully selected for each dataset. Additionally, if different objects are close to each other, then the method matches them as the same. Vice versa, if an object changes its position and the distance between positions in the catalogues is greater than the threshold then the objects are not matched.

Machine learning algorithms, on the other hand, can process a large volume of data, work with sufficiently noisy data, can be based on different features, discover dependencies in data. Most of machine learning algorithms have better results than ε-neighborhood method on matching catalogs containing nearby different objects or noisy data. A number of ways exists to improve the quality of model trained. Comparison with statistical methods, machine learning requires extensive preparation of train and test datasets and subsequent model training. A model developer has to have in mind overfitting problem also.

The decision tree and algorithms based on gradient boosting showed for datasets containing large number of "bad" or noisy data. All presented quality metric values are higher than 70% even for the third case. This shows that the algorithms found the majority of correct matches and reject matching for different objects. In our experiments, the decision tree and XGBoost performed slightly better.

The choice of specific method for cross-matching of arbitrary astronomical catalogues depends on these catalogues. The choice can be based on experiments applying different machine learning methods for data from these catalogues. An ensemble of machine learning methods can also be applied.

5   Conclusions

The paper investigates the problem of cross-matching astronomical catalogues applying machine learning methods. An overview of recent existing approaches to cross-matching is presented. A generalized approach to solve the problem is proposed. Depending on the quality of data in matched catalogues three different cases are distinguished and the respective experiments applying the approach are performed. The conventional ε-neighborhood method is used as the baseline within the experiments. When comparing the results, the advantages and disadvantages of applied methods are considered.

To summarize, the choice of the method to be used for cross-matching astronomical catalogues depends on the quality of the data, their volume, and available computation facilities. If the matched catalogues are trusted, clean, data are observed at the same epoch, then the ε-neighborhood method or similar one can be applied. However, if the catalogues contain nearby different objects, or objects with high proper motion observed at different epochs, then machine learning methods should be applied. In this case, a specific attention should be paid to feature selection, creation of the train and the test sets, and training the model.

Acknowledgements. This work uses data from the European Space Agency (ESA) Gaia mission (https://www.cosmos.esa.int/gaia) processed by the Gaia Data Processing and Analysis Consortium (DPAC, https://www.cosmos.esa.int/web/gaia/dpac/consortium). Funding for DPAC was provided by national institutions, in particular institutions participating in the Gaia Multilateral Agreement.

The work is financially supported by the Russian Foundation for Basic Research, project 19-07-01198.

References

1. Hubble Space Telescope. https://hubblesite.org/. Accessed 20 Mar 2021
2. Gaia Archive at ESA. https://gea.esac.esa.int/archive/. Accessed 20 Mar 2021
3. Infrared Astronomical Satellite (IRAS). https://irsa.ipac.caltech.edu/Missions/iras.html. Accessed 20 Mar 2021
4. Krakowski, T., et al.: Machine-learning identification of galaxies in the WISE SuperCOSMOS all-sky catalogue. Astron. Astrophy. **596**, A39 (2016)
5. Wide-field Infrared Survey Explorer (WISE) and NEOWISE. https://irsa.ipac.caltech.edu/Missions/wise.html. Accessed 20 Mar 2021
6. SuperCOSMOS Science Archive. http://ssa.roe.ac.uk/WISExSCOS. Accessed 20 Mar 2021
7. Marrese, P.M., et al.: Gaia data release 2 - cross-match with external catalogues: algorithms and results. Astron. Astrophys. **621**, A144 (2019)
8. Marrese, P.M., et al.: Gaia data release 1 - cross-match with external catalogues: algorithms and results. Astron. Astrophys. **607**, A105 (2017)
9. The Wide-field Infrared Survey Explorer at IPAC. The AllWISE Data Release November 13, 2013. https://wise2.ipac.caltech.edu/docs/release/allwise/. Accessed 20 Mar 2021
10. The Ninth SDSS Data Release (DR9). https://www.sdss3.org/dr9/. Accessed 20 Mar 2021
11. Rohde, D.J., et al.: Applying machine learning to catalogue matching in astrophysics. Mon. Not. R. Astron. Soc. **360**(1), 69–75 (2005)
12. Ogle, P.M., et al: Rule-based cross-matching of very large catalogues in NED. arXiv preprint arXiv:1503.01184 (2015)
13. Shi, X., Budavari, T., Basu, A.: Probabilistic Cross-identification of Multiple Catalogues in Crowded Fields. Astrophys. J. **870**, 51 (2019)
14. Budavari, T., Basu, A.: Probabilistic cross-identification in crowded fields as an assignment problem. Astrophys. J. **152**, 86 (2016)
15. Munkres, J.: Algorithms for the assignment and transportation problems. J. Soc. Ind. Appl. Math. **5**, 32–38 (1957)
16. Soodla J.: A System for cross-matching all-sky surveys. Master thesis, Uppsala Universitet (2019)
17. Akhmetov, V., Khlamov, S., Dmytrenko, A.: Fast coordinate cross-match tool for large astronomical catalogue. In: Shakhovska, N., Medykovskyy, M.O. (eds.) CSIT 2018. AISC, vol. 871, pp. 3–16. Springer, Cham (2019). https://doi.org/10.1007/978-3-030-01069-0_1
18. Scikit-learn: Machine Learning in Python. https://scikit-learn.org/stable/index.html. Accessed 20 Mar 2021
19. Lindegren, L., et al.: Gaia data release 2 the astrometric solution. Astron. Astrophys. **616**, A2 (2018)

Pipeline for Detection of Transient Objects in Optical Surveys

Nicolai Pankov[1,2]($\boxtimes$) , Alexei Pozanenko[1,2] , Vladimir Kouprianov[3,4] ,
and Sergey Belkin[1,2]

[1] National Research University Higher School of Economics,
Staraya Basmannaya 21/4, Moscow 105066, Russia
nspankov@edu.hse.ru
[2] Space Research Institute of the Russian Academy of Sciences,
Profsoyuznaya 84/32, Moscow 117997, Russia
[3] University of North Carolina at Chapel Hill, 310 South Road,
Chapel Hill, NC 27514, USA
[4] The Central Astronomical Observatory of the Russian Academy of Sciences
at Pulkovo, Pulkovskoye Chaussee 65, Saint-Petersburg 196140, Russia

Abstract. Identification and following study of optical transients (OTs)
associated with cosmic gamma-ray bursts (GRBs) and gravitational wave
(GWs) events is a relevant research problem of multi-messenger astronomy. Since their first discovery, one of the greatest challenges is the
localisation uncertainty. The sources of OTs are initially localised with
space gamma and X-ray telescopes or ground-based laser interferometers
LIGO, Virgo and KAGRA having the poor positional accuracy on average. A joint localisation area typically covers about 1000 deg^2 of the sky
based on previous runs of LIGO and Virgo. The last 25 years has seen
a rapid development of the robotic optical surveys. Such instruments
equipped with wide-field cameras allow to cover the entire localisation
area in several scans. As the result, a massive amount of scientific products is generated, including bulky series of astronomical images. After
their processing, large object catalogues that may contain up to 10^5 celestial objects are created. It is necessary to identify the peculiar objects
among other in the formed catalogues. Both data processing and identification of OTs must be carried out in real-time due to steep decay
of brightness. To response pointed problems, the software pipelines are
becoming a relevant solution. This paper provides a complete overview
of the units of the actively developed pipeline for OT detection. The
accuracy and performance metrics of the pipeline units, estimated for
two wide-field telescopes are given. In conclusions, the future plans for
the development are briefly discussed.

Keywords: Optical transients · LIGO · Virgo · KAGRA · Kilonovae ·
Gamma-ray bursts · Real-time image processing

© Springer Nature Switzerland AG 2022
A. Pozanenko et al. (Eds.): DAMDID/RCDL 2021, CCIS 1620, pp. 104–134, 2022.
https://doi.org/10.1007/978-3-031-12285-9_7

1   Introduction

Historically, the first detections [1,3] of gamma-ray bursts were taken in 1960–70s by the Vela satellites. Unfortunately, their instruments had a limited sensitivity to provide the accurate location of the gamma-ray or X-ray sources. Additionally, a lack of fast communication between spaceships and telescopes on Earth has not allowed to follow-up GRBs and find their optical sources. To improve the situation, the NASA launched the Compton Gamma-Ray Observatory (CGRO) spacecraft on April 5, 1991, which mission was aimed to study the Universe at high energies. The Burst And Transient Source Experiment (BATSE) on board of the CGRO served as the all-sky monitor in range of energies 20 kev – 8 Mev and detected about 2,700 GRBs during the operational period. Specially for the CGRO, the BACODINE coordinates distribution network was designed, and when the CGRO has been deorbited, it was renamed to the Gamma-ray Coordinates Network GCN[1], which is still widely used today. The analysis of the BATSE bursts showed important facts about GRBs nature, including their isotropic distribution across the sky and a phenomenological classification by a duration [5]: short ($T_{90} \lesssim 2$ s) and long ($T_{90} \gtrsim 2$ s) events. Later, with new space observatories, the spectral parameters [56,57] were also included into this classification scheme. The growning GRB science proposed the theory of expanding fireball [52] as the GRB emission model. On of the most important implications leads to the existence of soft GRB radiation at X-rays, in optical and in radio-waves, but no credible one has been observerd yet. A great contribution to multimessenger astronomy was made by the discovery of GRB 970228 [32] on February 28, 1997 initially in gamma-rays and more precisely, with arcminute resolution, in X-rays by BeppoSAX satellite, equipped with the wide-field instrument. After the coordinates were measured, the ground-based optical telescopes were pointed to the sky error box. With the grateful help of Roque de Los Muchachos Observatory in La Palma, first optical images of GRB 970228 were taken on February 28 and March 3, 1997. Later, by observing GRBs afterglow from X-ray to radio, astronomers proved the linkage between long GRBs and SNe [6,7], that was firstly mentioned by [54,55]. Their spectroscopy reveals that GRBs are extragalactic sources. The most distant GRB 090423 [33] and GRB 090429B [35] are at spectroscopic redshift $z \approx 8.2$ [34] and photometric[2] redshift $z \approx 9.4$ [36], respectively. This decade, several attempts has been made to expand the wavelength coverage for the purpose of multiwavelength exploration of the Universe. The Neil Gehrels *Swift* Observatory [47] with the Burst Alert Telescope (BAT) [28,48], the X-Ray Telescope (XRT) [50] and the Ultraviolet/Optical Telescope (UVOT) [49] on board was launched in 2004 and opened the new 'Swift' era for discoveries in investigation of the gamma-ray bursts, thanks to the fast slewing of the spacecraft, great positional accuracy (from arcminutes for the BAT to arcseconds for the XRT and UVOT) and real-time alerting to the GCN. The Swift GRBs sample reveals numerous interesting features in their light curves. The *Fermi* (former GLAST) [29] observatory was

---

[1] https://gcn.gsfc.nasa.gov/gcn3_archive.html.

[2] A technique to estimate a redshift by constructing a SED from imaging observations.

launched on 11 June 2008 and equipped with pair conversion Large Area Telescope (LAT), and Gamma-ray Burst Monitor (GBM) instruments. The Fermi LAT is capable to detect the neutrino emission of the GRBs at TeV energies with under 1 deg² accuracy; while the GBM is sensitive to gamma-rays with energies up to 30 MeV, but has much poorer precision (100 deg²). In the present time, multiwavelength astronomy actively increases the number of GW observations. The main objects of study are coalescing binaries such as binary neutron star (BNS) mergers. The theory proposes that coalescing binaries can produce gravitational radiation ('chirp' signal) and short GRBs after collision [2,58,64,65,68]. However, construction of gravitational wave detectors began in early 1990-s from projects LIGO [66] and Virgo [67]). The measurement method is used to receive gravitational signal relies on the detection of the gravitational strain h_c at Earth. The strain amplitude is given by the formula (1) [64]:

$$h_c \propto 10^{-23} h_0 \mu_{\mathrm{NS}}{}^{1/2} M_{\mathrm{NS}}{}^{1/3} f^{1/6} \frac{r_H x^{5/6}}{D_L} \tag{1}$$

where h_0 is scaling amplitude, μ_{NS} is reduced BNS mass, M_{NS} is mass of the NS, f is knee frequency of the gravitational wave, r_H is Hubble radius, D_L is luminosity distance and $z = x - 1$ is redshift. To observe the gravitational waves from the merging binaries, the sensitivity for the detection rate about 1 event per year must be [64] $\sim 10^{-21.6} h_0$, which is limited by the noise. As of year 2017, in cycle O2 of gravitational-wave observations, *Advanced LIGO* [69]) and *Advanced Virgo* [70] achieved the best strain sensitivity of $5 \times 10^{-24} \sqrt{\mathrm{Hz}}^{-1}$, enough to detect the BNS merger GW170817 [44] associated with short GRB 170817A [44] detected by the *Fermi-GBM* [72] and the *INTEGRAL* [71] spacecrafts with a Shapiro delay [45] of about 1.7 s, that proves an effect of general relativity. The accompanning optical transient (kilonova) AT2017gfo [31] was discovered by the *Swope Supernova Survey* since 11 h of hardworking observations after GRB and located at $z = 0.00968$ [73] (equivalent to the distance of ~ 40 Mpc) at a projected distance of ~ 2 kpc from the centre of the host galaxy NGC 4993. The GW170817 projenitor was the BNS merger with the resulting mass $\mu \approx 2.82^{+0.47}_{-0.09} M_\odot$ [44]. This particular multiwavelength observation is unique in its kind: an only GW event with a known optical source up to date. It is clear that a large locaisation uncertainty (28 deg² [74]), and a steep power law decay of brightness is an impassible wall for a second AT2017gfo like event. These problems requires the usage of the robotic wide-field surveys capable to cover the entire localisation area. An integral and extremely important part of such facilities are the software pipelines that control telescope marshalls, process the survey data and identify the OTs in real time. The following section of the paper provides the literature review of successfully operating transient facilities and their software for processing the data and OT identification.

2   Literature Review

Transient factories are complex combinations of robotic optical telescopes remotely or autonomously operated, equipped with a wide-field survey camera and an infrastructure for storing, processing and transferring scientific data. It is important to reveal many aspects of the software already has been developed for processing the astronomical data and detection of OT's by other teams before creating own solution.

2.1   ROTSE

One of the first such facilities is the Robotic Optical Transient Search Experiment (ROTSE) that was launched in the late 1990-s. It is designed to automatically follow-up optical counterparts of GRBs following the alert from a space observatory with a cadence of several seconds. The ROTSE was presented in three generations of telescopes: the ROTSE-I [30], the ROTSE-II [30] and the ROTSE-III [37] global network. As said in [37], the ROTSE software runs in the Linux environment and performs the image calibration, the object extraction, the astrometric and photometric reduction, and finally OT detection and their classification. An image scalibration step is followed through standard procedures of dark-frame subtraction and flat-fielding. An object extraction is implemented via SExtractor [21] software. The extraction method is based on the signal-to-noise ratio (SNR) thresholding, pixel labelling, and finally connection of labeled pixels (objects). The SExtractor also provides the fixed aperture photometry (5 px diameter) on them. The ROTSE astrometric reduction is tied to the stars from the Hipparcos and USNO catalogues, the Landolt or USNO-A2.0 standard stars are used for the photometric reduction. OTs are detected using the images substraction technique and cross-matching with the reference catalogues (Hipparcos or USNO) or between local catalogs from sequential observations. To eliminate the hot pixels and cosmic ray tracks, the telescopes shift their aspect from frame to frame. Also, from the list of detected objects known asteroids or masking stars can be flagged by the ROTSE software. The ROTSE autonomous data acquisition system daq is consisted of several daemons communicating with via the shared memory. An alert daemon establishes a connection with GCN via TCP/IP sockets. Other daemons are used for communication with camera, weather station, user, etc. The ROTSE-I has taken the first ever imaging of the optical prompt emission of GRB 990123 [39] ~25 s after BATSE trigger [4,38]. It is worthly note that results obtained have showed that transient factories are suitable for operating in near real-time mode to follow-up GRB counterparts in optical, which would be impossible without the reliable software. The given fact boosted many teams around the world to create the transient farms.

2.2   ZTF

The Zwicky Transient Facility (ZTF) [9] was launched in 2017 based on the P48 telescope equipped with a wide-field (FoV = 47 deg^2) camera. The survey repeatedly covers the entire visible part of the northern sky. The ZTF uses the Zwicky Science Data System (ZSDS) deployed at IPAC [10,11] and runs the data processing pipelines, file and database handling, workload monitoring and alerting with Kafka. According to [10,11], the image prcosessing pipeline has several stages and starts with raw image calibration. At this stage, the raw quandrant images are unpacked and corrected for the bias-overscans, thereafter, high-frequency flat-fielding performed. Several filters applied to reject the out-lied pixels and to mask bad pixels. Additionally, as given in [11], many other routines are performed at the current step: detector non-linearity correction, masking the haloes from the bright stars, aicraft and satellite streaks, and cosmic ray elimination. For masking the tracks from aicrafts and satellites, authors [11] use the `CreateTrackImage` [82] software, which recognizes them as long and narrow blobs of pixels. Cosmic ray elimination is done further when stacking the image series into one by $n\sigma$ averaging the pixel stack [11]. Then, the `SExtractor` is used to form the initial catalog of objects and for aperture photometry on them. After that, the `DAOPHOT` [41] derives the PSFs and produces the PSF-photometry by fitting the model function into sources. At the next, astrometric reduction is performed with the `SCAMP` against the Gaia DR2 [40] not saturated stars satisfying $12 \le G \le 18$. The `SCAMP` [13] cross-corellates projected coordinate pairs of the objects extracted from the image and written to FITS LDAC files by `SExtractor` with the reference catalog twice. After the first run the astrometric solution is computed and it provides the FOV pointing, rotation, and pixel scale at the image center. The second run is required to refine the previous astrometric solution with a new one using fourth-order polynomial represented in TPV convention [11]. Sequentially, the photometric reduction is performed against standard stars from filtered and prepartitioned PanSTARRS PS1 [26] catalog to calculate apparent magnitudes. The photometric solution is retrieved from the fit of $m_{diff} = ZP_f + c_f PS1_{clr}$, where $PS1_{clr} = g_{PS1} - r_{PS1}$, $r_{PS1} - i_{PS1}$ are calibration star colors, ZP is the photometric zero-point, and c_f is the color term for every ZTF filter $f = g, r, i$. ZTF uses image difference technique to find candidate transients: the survey images are substracted from the reference template images (produced earlier) with the `ZOGY` [84]. The classification of defects on the residual images is performed with convolutional neural network `BRAAI` [75]. Finally, the candidate transients to GWs/GRBs are found and identified with `ZTFReST` [60] pipeline. The recent interesting results of the ZTF survey are detections of 'orphans' [61,62,62]. Orphans are potential candidates to optical components of sGRBs from BNS or NSBH mergers. The GRB jets collimated into narrow angles, while their afterglow could be observed at wider angles after the jet-break $\theta \propto \Gamma^{-1}$, where Γ^{-1} is the bulky Lorentz factor [51,63]. Orphans afterglows are not associated with GRB prompt emission [63], but further observations might indicate the presence of X-ray and/or gamma-ray components. The ZTF experience will be applied in next-gen Large Synoptic Survey Telescope (LSST)[3] project at Vera Rubin's Observatory being able to catch faint transients down to 23 magnitude.

---

[3] https://www.lsst.org.

2.3   MASTER-Net

The Mobile Astronomical System of TElescope Robots Network (MASTER-Net) [8] was launched in 2002 and consists of several telescopes in Russia, one on Canary Islands, one in South Africa and one in Argentina. Each 0.4-meter telescope equipped with two ultra wide-field cameras with a maximal field of view of 1,000 deg^2. Photometry is performed in 4 optical bands: B, V, R and I, or without installed filters in clear light. Typically with a positioning rate of 30°/s, the sky is scanned during a single night, reaching a limiting magnitude of 19 . . . 20. The MASTER-Net's image processing pipeline provides similar structure in comparison with previously reviewed solutions. As mentioned in [8,83] the prcosessing starts with image calibration for bias, dark and flat. The MASTER-Net pipeline detects and classifies objects with SExtractor [83]. The aperture photometry on the objects is performed with phot package of IRAF [43]. For obtaining the astrometric solution the imcoords package of IRAF is used as said by [43]. It provides two algorithms tolerance and triangles for the astrometric reduction. The first one is simple coordinate pair matching of image objects with catalog ones within a certain tolerance. The second one is more effective and complex, finds matchings of triangular patterns in coordinates between image objects and catalog entries. Both from [8] and [43] may be deduced that astrometry is done against standard stars of Tycho2, USNO-B1.0, and SDSS-DR7 [25] catalogs. By revealing the same references, in the process of photometric reduction, differences of instrumental magnitudes and catalog magnitudes are minimized for the given image. Calibration stars are selected from previously pointed catalogs and additionally with Landolt stars. In the case of SDSS-DR7, to calibrate magnitudes measured in B, V, R and I bands, Lupton[4] transformations are applied [43]. As can be seen from [43], the strategy of search for transients initially relies on automatic coordinate matching with reference catalogs (indicated above). It provides three categories of identification: 'the known stars', the objects matched by coordinates and magnitudes, 'flares', the objects matched by coordinates but with significant excess of brightness compared with catalog values, and 'unknown', potential candidate transients. Further revealing of [43] has showed, that the approach of a personal inspection and analysis is involved. A person analyses light curves and archived images for object existence and its state. Also, additional search for the objects inside the circle of 5 arcsec radius is performed via Vizier[5]. The MASTER-Net regulary scans error boxes of GRBs and provides astronomers an ability to image prompt optical phase of GRBs; for example, there are several results published in the following paper [42].

2.4   IKI GRB Follow-Up Network

The IKI GRB Follow-Up Network (IKI-GRB-FuN or GRB-IKI-FuN, interchangeably) connects several observatories in both hemispheres. Almost each

---

[4] https://classic.sdss.org/dr7/algorithms/sdssUBVRITransform.html.
[5] https://vizier.u-strasbg.fr/viz-bin/VizieR.

observatory equipped with telescopes (including wide-field ones with $FoV \geq 6.15$ deg) that unique in terms of the configuration of the detectors and optics used. An aperture size of the telescopes varies from 0.36 (RC-36 of Kitab-ISON observatory) to 11 m (SALT of South African Astronomical observatory). For photometric observations there are g, r, i, z (AB system) and B, V, R, I, J, H, K (Vega system) passbands available, depending on the observatory equipment. The climate and seeing conditions may differ from one observatory to another. The geographical locations of the observatories of the IKI GRB FuN are placed onto the Earth map and shown in the Fig. 1.

Fig. 1. The IKI GRB Follow-Up Network: the observatories are coded with stars.

On behalf of IKI-GRB-FuN team, we observe and study optical transients such as GRB afterglows, supernovae, kilonovae and tidal disruption events. In order to process the observational data from the various telescopes, search and identification of optical transients we initiated the development of the image processing pipeline (see more in Sect. 3).

2.5   Another Surveys

In fact, it is necessary to mention that there are many more such surveys (order does not reflect the importance or preference) as the Burst Observer and Optical Transient Exploring System (BOOTES) [14], the Gravitational-wave Optical Transient Observer (GOTO) [15], the Dark Energy Camera (DECam) [77], the Black GEM [78], the All Sky Automated Survey for Supernovae (ASAS-SN) [53], and many others. Unfortunately, the overview of these ones does not fit within this paper.

3   Image Processing Pipeline

3.1   Motivation

Studying of short OTs is still a difficult problem for the project around the world. This explains, why so far, there is only one KN associated with GW/GRB has been discovered. However, the estimated rate of joint detections of LIGO and Fermi GBM GWs/GRBs is up to 3.90 yr^{-1} [76]. So the chances of new discoveries are far from zero. Our motivation is quite simple and based on the experience of the AT2017gfo campaign [31,79]: the software pipelines are necessary to solve the problem, but no specific solution is distributed publically.

3.2 Requirements

The main requirement for the processing pipeline is that operation must be carried out in close to real-time (during an exposure time). To satisfy this requirement, the processing operation must be well optimized. Along with this, the processing must be performed in a stream mode for a single exposure image and immediate result. When the amount of generated data is large for immediate processing or high SNR is needed, batch processing of series of images may take place. A great advantage of the batch processing is a possibility for scheduling while a telescope is operated in scanning mode. This strategy can improve the efficiency of the resources used. The another requirement is the customization for a particular telescope of the IKI GRB FuN. Also, all the processing routines must meet the requirements of accuracy and reliability. Additionally, the false alert rate must be as lowest as possible.

3.3 Architecture

The current pipeline is implemented as a self-consistent monolithic module (i.e. all tasks are performed inside the module, no external applications required) consisting of the several units (submodules) that solve the separate tasks. A combination of tasks is the core of functionality customization of the pipeline. Each task can have extensions which are implemented as plugins. The main advantage of this idea is to be able to customize a particular task of the pipeline and easily implement new ones. Settings for a given submodule or a plugin are written in the configuration file. There is an ability to switch between the configuration files which is relevant to tune the settings for the telescopes and embed the pipeline to existing or planning surveys. A schematic diagram of the pipeline architecture is shown in the Fig. 2.

3.4 Implementation

The algorithmic core of the pipeline is the powerful python package APEX [16,17] for astronomical image processing. APEX provides image calibration, extraction and measurement of objects with both aperture and PSF-fitting, deblending of visual-multiple systems and accurate astrometric processing with several algorithms. A plugin system, shown previously is already implemented in APEX with the API of the BasePlugin base class. Thus, to match the core the chosen development environment is CPython 3.8.12. Its allows one to elegantly refactor and upgrade existing core package with new features and write scripts, link

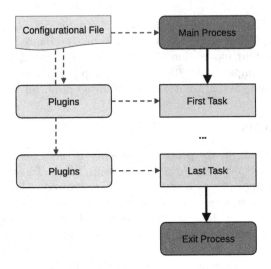

Fig. 2. A plugin architecture of the pipeline.

external DLL's to the python C-API or compile extensions written in C/C++ or Fortran with `Cython` or `numpy.distutils`. The image processing from start up to the identification of transient candidates is performed according to the flow chart in the Fig. 3.

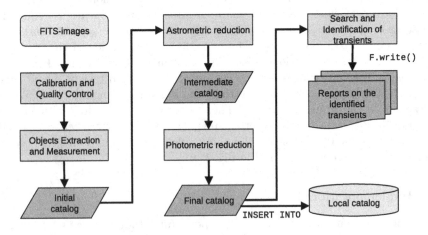

Fig. 3. The image processing flow chart.

The organization of the pipeline units is comparable to that from the ZTF or the TERAPIX pipeline [12].

Calibration and Quality Control. The raw uncalibrated images are prone to defects and errors, which decrease the overall processing quality. It is very important to drop the artifacted images and filter out the following defects when possible:

1. strong background gradient across the image field throughout the series of images, associated with a background gradient for objects at high airmass or with a bright star outside the field of view but close to it;
2. strong background deviation across the single image with comparison to other images in series, which results in decreasing of the limiting magnitude and a small number of detected objects, caused by:
   (a) short flashes inside a telescope dome (leakage of light through the optical path or random glares from the hardware LEDs);
   (b) clouds or the Moon light that temporarily come into the field of view;
3. significant object ellipticity or deformation in the field of view due to bad telescope tracking:
   (a) in one image in series;
   (b) in several images in series;
4. long trails across the entire image diagonal caused by satellites;
5. 'ghosts' – false images that may be caused by:
   (a) residual charge on the CCD after previous observations of the current or another field;
   (b) a sudden shift of the field of view during single exposure;
6. hot (bad) pixels – pixels affected by dark current or simply broken;
7. cosmic ray (CR) tracks – formations elongated by several pixels with sharp edges, like hot pixels; caused by penetration of the CCD with charged muons; appear in different places in the image; the higher the altitude of the observatory and the longer the exposure of the image, the greater the probability of the appearance of CR tracks.

Thus, calibration and quality control are split into several steps:

1. raw image calibration;
2. quality control;
3. stacking.

We used to apply quality control at the beginning of processing, then placed it before stacking. This allowed us to get rid of the problem with processing the raw images. Also, we not here that the correction of the hot pixels and CR tracks are performed at the calibration step.

Raw Image Calibration. A calibration procedure performs not only bias and dark frames subtraction, and flat-field correction, but also the cosmetic correction of the hot pixels and the cosmic ray (CRs) and executed via the `apex_calibrate.py` from APEX. Bias or dark frames are found automatically by scanning the specified calibration directory `calib_dir` or the sub-folders in the current directory. Appropriate bias and dark frames are chosen by the frame type, detector temperature and exposure time from the image headers (keys EXPTYPE, CCD_TEMP and EXPTIME, respectively, by default). Bad pixels are eliminated by replacing them with averages/medians across their neighborhood. The coordinates of the bad pixels or the bad columns/rows can be provided after visual inspection by hand and written to the pipeline configuration file or detected automatically with sigma clipping. The CR tracks are eliminated from the images with Astroscrappy[6] detect_cosmics function implementing the L.A.Cosmic [18] algorithm and called via more convenient APEX wrapper. A weighted median filter is used to minimize background gradient across the image field and implemented as clean_estimator plugin in APEX. The resulting background map is subtracted from the initial image.

Quality Control. Quality control is implemented as script apex_qc.py that is executed after calibration step. Detection of defects relies on the *sigma clipping* algorithm and implemented inside apex_qc.py script. A sigma clipping acts iteratively on a sample of a parameter and marks significant outliers ($>k\sigma$) from the mean at each step, where, σ – standard deviation of a sample and k is a real number. The quality control parameters are followed:

1. backlevel – mean background level in the image;
2. ndet – number of detected objects in the image;
3. ellipticity – mean ellipticity of objects in the image;
4. fwhm – image seeing in pixels.

Long trails are detected as objects with large ellipticity: at least one of axes is comparable to the image diagonal. The images with significant parameter outliers are considered artefacts and do not pass trough the quality control.

Image Stacking. To increase the signal-to-noise (SNR) ratio, a series of images can be deep stacked. The supported stacking modes are a difference, an average, a median and a sum. Before the stacking, an alignment must be performed, because objects' positions are often shifted between the images over the exposure time. The first and most reliable alignment method is based on cross-matching objects between the reference image and other images in series. Shifts are calculated by solving a system of Eqs. 2 in which the coeffitions A, D (shifts along the X- and Y-axes) and B, C, E, F (elements of the rotation-scale-skew-flip matrix) are affine transformations which are necessary to align one image with another.

$$X_p = A + BX_m + CY_m \tag{2}$$
$$Y_p = D + EX_m + FY_m \tag{3}$$

where X_p, Y_p are vectors of (X, Y) coordinates of the objects in the given image (plate coordinates) and X_m, Y_m are coordinates in the reference image ('measured' coordinates). In the alignment routine, the cross-matching is performed by

---

[6] https://github.com/astropy/astroscrappy.

the algorithms described below (see 'astrometric reduction' for details). Another alignment method uses hour angle (HA) and declination (DEC) tracking rates from the image headers (keys HA_RATE and DEC_RATE, respectively). Then HA and Dec rates are converted to appropriate X- and Y-coordinate shifts, shift is simply $HA(Dec)_{rate} \times t_{exp} \times scale$, where t_{exp} – exposure time. This method is applicable only if the astrometric solution is already performed for all images in the series and is much faster than the first one but has a great disadvantage – small uncertainties in tracking rates lead to poor alignment.

Initial Astrometry. Most of the survey images obtained during observations over time do not have an astrometric binding, which is strictly needed for further reduction routines. For example, a famous among the astronomers open source project and service astrometry.net[7] [27] provides an API for the remote blind astrometric reduction and can be deployed on the local machine. The astrometry.net detects objects in a query image and seeks its own catalog of appropriate *index FITS files, not regular images,* containing quads of stars (asterisms) that already astrometrically binded, for appropriate ones. Matched asterims are used to calculate necessary astrometric parameters: plate scale and the central pixel celestial coordinates. We implemented the wcs_hedit.py ultimately simple script to construct the initial astrometry parameters from an image header (a CCD/CMOS pixel size and a telescope effective focal length (EFFL), to identify the image field in the further astrometric reduction. However, this procedure is used in SCAMP as the first step of the astrometric reduction. The formulae (4) represent the initial astrometry structure.

$$
\begin{cases}
\text{CRPIX1} = [0.5 \times \text{WIDTH}] \\
\text{CRPIX2} = [0.5 \times \text{HEIGHT}] \\
\text{CRVAL1} = \text{RA\_CENTER} \\
\text{CRVAL2} = \text{DEC\_CENTER} \\
\text{SCALE} = \frac{\text{PIXSIZE}}{\text{FOCALLEN}} \\
\text{CD} = \begin{pmatrix} -\text{SCALE} & 0 \\ 0 & \text{SCALE} \end{pmatrix}
\end{cases}
\tag{4}
$$

where CRPIX1,2 are plate coordinates of the reference pixel (whole numbers), CRVAL1,2 are celestial coordinates of the reference pixel, given by the approximate numbers RA_CENTER, DEC_CENTER, provided with accuracy of field of view of a given image, SCALE is an average plate scale (arcsecond/pix), CD is a matrix of both increment and rotation of World Coordinate System (WCS) about the image coordinate system at the reference pixel, WIDTH and HEIGHT are the image width and height, respectively. All these parameters are written to the image header.

---

[7] http://astrometry.net.

3.5   Objects Extraction and Measurement

An automatic objects extraction is performed to create the catalog of objects, that will be feed to the further processing routines. A task is split into several steps

1. segmentation – image binarization, where background pixels are 0 and objects pixels are 1;
2. filtering the binary image to increase the detectability of the faint objects;
3. connection pixels into labeled groups;
4. deblending overlapping objects (in visual multiple systems);
5. measurement of the objects.

Segmentation. Segmentation is performed based on a global threshold algorithm after the background map is subtracted and given by the expression [16,17]:

$$\begin{cases} \hat{\mathbf{M}} = 1 \ \text{if} \ \hat{\mathbf{I}} > B + k\sigma \\ 0, \ \text{otherwise} \end{cases} \tag{5}$$

where $\hat{\mathbf{M}}$ is binary 2-D mask, $\hat{\mathbf{I}}$ is input image, B is mean background level and $k\sigma$ is signal-to-noise ratio (SNR) in units of background standard deviation (σ). B and σ are computed as the mean and standard deviation of an image in the sigma-clipping, where 3σ outliers are rejected iteratively. Then B and σ are refined by fitting the 2D Gaussian (6) to the image histogram [16,17].

$$I(x,y) = \frac{1}{\sqrt{2\pi\sigma^2}} \exp \frac{-(B-I)}{2\sigma} \tag{6}$$

The fragment of the image and its binary mask are presented in the Fig. 4.

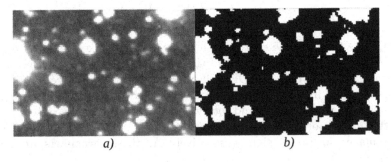

a)   *b)*

Fig. 4. The result of the image binarization: (a) the fragment of the original image and (b) the same fragment after binarization.

Filtering. Faint object images at low SNR ($k < 3$) are clogged with noise pixels. The morphological filter [16,17] used in APEX, eliminates them and improves the detectability of faint objects. The filter kernel of size N × N is given by the expression 7

$$\begin{cases} \hat{\mathbf{M}} = 1 \ \text{if} \ \sqrt{(i - [N/2])^2 + (j - [N/2])^2} \leq [N/2]^2 \\ 0, \ \text{otherwise} \end{cases} \tag{7}$$

where $i, j = 0, 1, \ldots, N - 1, [\ ]$ is integer part of the number, and N is chosen so that the filter kernel covers the object image. Filter action on the image is expressed by the formula 8

$$\begin{cases} \hat{\mathbf{M}}(x,y) \to 1 \ \text{if} \ M \oplus K(x,y) \geq d \sum_{n,j=0}^{N-1} K_{ij} \\ 0, \ \text{otherwise} \end{cases} \tag{8}$$

where d is empirical threshold value, usually $d \approx 0.1$. The result of the filtering is shown in the Fig. 5.

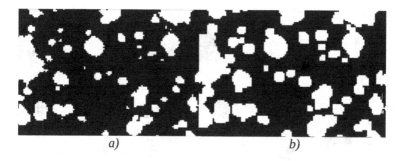

a)   b)

Fig. 5. The result of the morphological filtering: (a) the fragment of the original image after applying the threshold algorithm and (b) the same fragment after filtering. Adopted from the paper [16]

Label Connected Pixels. Pixels (in other words components), extracted by a threshold algorithm, are combined into groups related to individual objects based on the connectivity properties of the group of pixels using the APEX wrapper around the implementation of the *Connected Components Labeling* (CCL) algorithm from the SciPy[8] library. Connectivity is the number of neighbors around the given pixel needed to determine the label. There are two values of connectivity are used in the pipeline: 4 and 8 (see Fig. 6).

---

[8] https://www.scipy.org/scipylib/index.html.

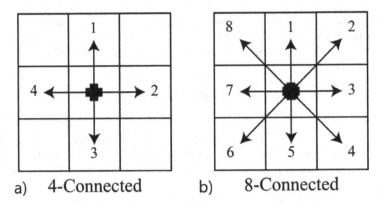

a)   4-Connected   b)   8-Connected

Fig. 6. Connectivity: 4-connectivity (a) checks only edge pixels, while 8-connectivity (b) checks also corner pixels. Adopted from [19].

The result of the labelling is shown in Fig. 7.

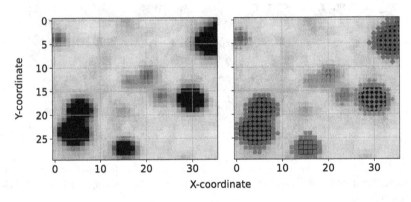

Fig. 7. The result of the labelling of the connected pixels: (a) the fragment of the original image (b) the same fragment after labelling. Individual groups are coded with different color and markers. (Color figure online)

Objects Deblending. Next stage of objects extraction is the resolving of overlapping images of close sources in visual multiple systems or *deblending*. For this purpose, a recursive implementation of the multi-threshold algorithm (MT) [20, 21] is used from APEX. The initial image of the blended system is re-thresholded in N levels higher than the original threshold. The more threshold levels are used, the better deblending, the optimal range is $N \sim 16 \ldots 32$ (balance of performance and ability for deblending). At each threshold level, the CCL algorithm is applied to label the pixels belonging to the individual components. The results of the deblending of the real image is shown in the Fig. 8.

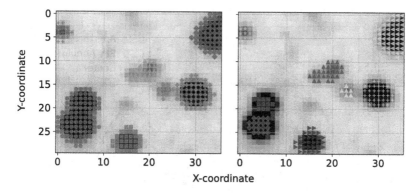

Fig. 8. The result of the deblending: (a) the fragment of the image with applied CCL, (b) the same fragment after deblending. Individual groups are coded with color, and their components denoted with different markers. (Color figure online)

Objects Measurement. After all objects were deblended, the isophotal analysis is performed to determine the isophotal positional parameters[9] of the objects: the center of mass, the rotation angle about the image coordinate system, and major and minor axes lengths[10]. Then the instrumental aperture photometry is performed. Aperture photometry – the measurement of the flux of an object within a given footprint, as shown in Fig. 9. The footprint shape and sizes are chosen with respect to the object FWHM and stored in the configuration file.

After the procedure of aperture photometry is finished, the aperture flux and flux error are obtained. Then, the point spread function (PSF) fitting is performed on the objects's pixels. The PSF fitting is mainly needed to find the accurate positions of the objects what is important for astrometry. Another cause is a multimode measurement of the blended groups of objects. The result of PSF fitting of the image of the blended group is shown in the Fig. 10.

There are several PSFs presented in the APEX standard library, but 2-D Gaussian is preferable (least number of parameters to fit), which is given by the formula

$$G(x, y) = G_0 e^{-\frac{x_{peak}^2}{2\sigma} - \frac{y_{peak}^2}{2\sigma}} \tag{9}$$

where G_0 is Gaussian amplitude, x_{peak}, y_{peak} are coordinates of the center mass of the object, and $\sigma \approx FWHM/2.355$ – a standard deviation of the object's PSF. Note, that positional parameters of the objects are recalculated after PSF fitting with comparison to the previous isophotal analysis. It is also important to mention that this is valid only for point-like objects. The PSF-fitting is performed by the usefull wrappers around the `SciPy MINPACKs` routines for the dumped least squares (Levenberg-Marquardt algorithm) fitting from `apex.math.fitting`

---

[9] In a similar way like SExtractor does, see the documentation: https://sextractor. readthedocs.io/en/latest/Position.html.

[10] All objects are assumed to be elliptical.

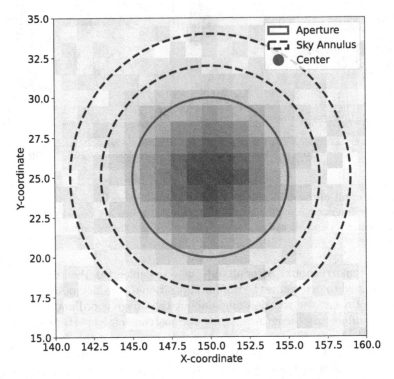

Fig. 9. Aperture photometry: annulus apertures are used for the local sky background measurement. Two blue dashed circles represent the inner and outter annuli, in turn. A green solid circle denotes the object aperture. The center of the object is indicated by a red dot. (Color figure online)

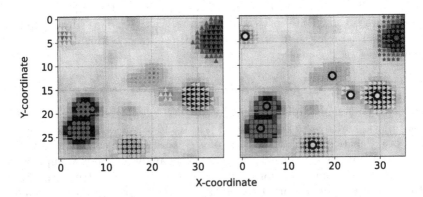

Fig. 10. The result of PSF-fitting of the objects: a) the fragment of the image after applied MT deblending, b) the same fragment after PSF-fitting. Bold dots denote the centers of mass of objects, and crosses denote PSFs. One tone color defines pixels belonging to the same blended group or the individual object. (Color figure online)

submodule. Both the aperture calculation and PSF fitting are multi-threaded routines what is available in the APEX standard library. A backend for the multi-threading is implemented via DASK[11] API.

3.6 Astrometric Reduction

The survey images are typically taken without any binding to an astrometric catalog. Thus, the objects positions are expressed only in the plate coordinates (X, Y). Conversion from the plate to celestial coordinates is required the astrometric reduction. The flow chart of the astrometric reduction unit is shown in the Fig. 11.

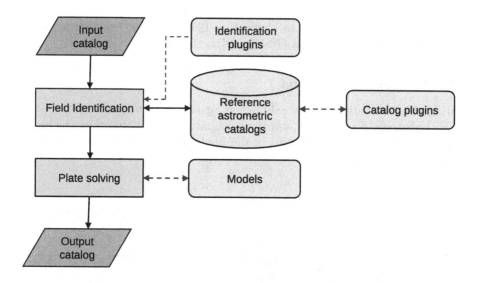

Fig. 11. The astrometric reduction flow chart

Initially, reference astrometric stars are selected from the list of all extracted objects. It is required for reference astrometric stars to have high SNR and small uncertainties in the centroid position for an accurate field identification. By cross-matching reference astrometric stars to the catalog, mapping from plate to celestial coordinates is calculated. There are several algorithms of automatic objects cross-matching presented in the APEX module: the triangular pairing [22], the distance-orientation algorithm[12]. All these algorithms modified in order

---

[11] DASK is a python library https://dask.org.
[12] See Kosik's paper http://www.iki.rssi.ru/seminar/virtual/kosik.doc.

to use a vote matrix [16,17,23]. Coordinates of reference astrometric stars in the image plate and in the catalog are tied and can be expressed by the model. To express the deviations from linear transformation of pixels to celestial coordinates, the SIP (simple image polynomials)[13] of $<3^{\mathrm{rd}}$ is used. The SIP model is given by the system of Eqs. (10)

$$\begin{pmatrix} x \\ y \end{pmatrix} = \begin{pmatrix} \mathrm{CD\_1\_1} & \mathrm{CD\_1\_2} \\ \mathrm{CD\_2\_1} & \mathrm{CD\_2\_2} \end{pmatrix} \begin{pmatrix} u + f(u,v) \\ v + g(u,v) \end{pmatrix} \tag{10}$$

where, x, y are celestial coordinates of the reference stars, and u, v are their plate coordinates. The solution of these equations provides mapping from plate to celestial coordinates for all objects in the field. There are many astrometric catalogs supported by APEX, but we decided to use local stored USNO-B1.0 [24] due to its all-sky coverage, sufficiently high astrometric accuracy (about 0.3 arcsec [24]) and completeness to about $R > 19.0$ (from our experience) unlike GAIA DR2 (complete to about $G > 17$ [40]).

3.7 Photometric Reduction

During the differential (i.e. by calibration against *reference photometric stars or RP*) photometric reduction, apparent magnitudes of all objects in the image field are computed. The flow chart of the photometric reduction unit is shown in the Fig. 12.

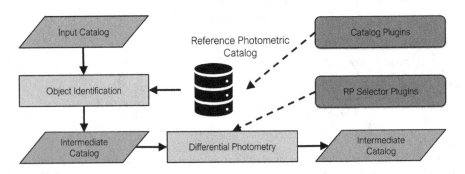

Fig. 12. The photometric reduction flow chart

Initially, the RP stars are selected from the list of all matched objects. We implemented a submodule `apex.photometry.usnob_rpselector` to obtain the RP stars from USNO-B1.0 catalog. Its flow chart is shown in the Fig. 13.

---

[13] Description of the SIP https://fits.gsfc.nasa.gov/registry/sip.html.

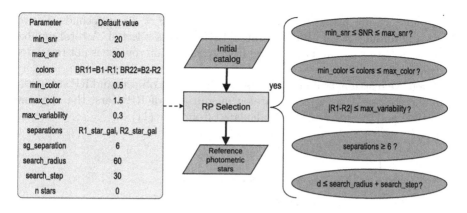

Fig. 13. The USNO-B1.0 RP selector flow chart.

The selection criteria based on the following *standard parameters* of stars:

- astronomical color index;
- variability;
- object type;
- SNR;
- angular separation;

An astronomical color index E is the magnitude difference between the two observational epochs in the same passband. Suitable colors are selected for the objects of interest. Appropriate values for GRBs afterglow $E(B-R) = 0.5 \ldots 1.5$. Color characterizes the spectrum of objects, which is difficult to obtain for faint objects. *A variability V* is the modulus of the magnitude difference between the 2 observational epochs in the same passband presented in the catalog. By restricting variability, it is possible for one to avoid the problem with the magnitude calibration in the long run observations (e.g. a several days or weeks). By default, the $V(|R1 - R2|) \leq 0.3$ is set. *An object type* is the classifier showing the degree of belonging of objects to point-like (i.e. stars) ones or galaxies. In USNO-B1.0 there are classifier per the filter per the epoch $B1s/g$, $B2s/g$, $R1s/g$, $R2s/g$ and Is/g are presented and expressed as whole numbers in range of 1 $\ldots 11$ from most galactic to most stellar order. We are using the classifiers with respect to the used catalog magnitudes: $B1s/g, B2s/g > 6$ for B1 and B2 magnitudes and $R1s/g, R2s/g > 6$ for R1, R2 and I magnitudes. The Is/g is not used because it is frequently broken or absent. *An angular separation* θ is the angular radius of the circle in the sky centered at the measured object, inside which the RP stars are selected, i.e. a search radius. If no RP stars for a particular object found within the circle, then the search radius is incremented with step $\Delta\theta$.

By default, we are using $\theta = 60$ arcseconds and $\Delta\theta = 30$ arcseconds. For an object a minimum of 1 and maximum of 4 RP stars is selected. A total number of RP stars N_{RP} that can be retrieved from the selection routine is not restricted by default ($N_{RP} = 0$).

Given the instrumental magnitudes of both regular objects and RP stars (and their uncertainties), as well as, catalog magnitudes of RP stars, the apparent magnitude and its uncertainty is calculated from Eqs. (11).

$$
\begin{cases}
m_{\text{diff}_i} = m_{\text{inst}_{\text{src}}} - m^i_{\text{inst}_{\text{std}}} + m^i_{\text{cat}_{\text{std}}} \\[2mm]
m_{\text{appar}_{\text{src}}} = \frac{1}{N}\sum_{i=1}^{N} m_{\text{diff}_i} \\[2mm]
\sigma_{\text{diff}} = \sqrt{\frac{\sum_{i=1}^{N}(m_{\text{diff}_i} - m_{\text{appar}_{\text{src}}})^2}{N-1}} \\[2mm]
\sigma_{\text{tot}} = \sqrt{\sum_{i=1}^{N}(\Delta m^i_{\text{inst}_{\text{std}}})^2 + (\sigma_{\text{diff}})^2 + (\Delta m_{\text{inst}_{\text{src}}})^2}
\end{cases}
\tag{11}
$$

where m_{diff_i} is differential object magnitude relative to the i-th RP star, $m_{\text{inst}_{\text{src}}}$ is instrumental magnitude of the object, $m^i_{\text{inst}_{\text{std}}}$ is instrumental magnitude of the i-th RP star, $m^i_{\text{cat}_{\text{std}}}$ is catalog magnitude of the i-th RP star, σ_{diff} is standard deviation of the differential magnitudes, $\Delta m^i_{\text{inst}_{\text{std}}}$ is absolute error of the instrumental magnitude of the i-th RP star, $\Delta m_{\text{inst}_{\text{src}}}$ is absolute error of the instrumental magnitude of the object, $m_{\text{appar}_{\text{src}}}$ is apparent magnitude of the object, σ_{tot} is total error of the apparent magnitude of the object.

3.8   Local Catalog of Objects

The pipeline creates the local catalog of the objects in form of SQLite[14] databases. Creation of databases and handling of queries is implemented by us as apex.util.sql submodule. We use sqlalchemy[15] python package, which provides Object-Relational Mapping (ORM) for manipulation of queries and data in object-orientated paradigm. In the current implementation of the pipeline we do not use a distributed database management system unlike in, for example, ZTF at IPAC, but a simply database per each processed image. Each database has a conventional naming according to the image name, and contains a table where each row represents an object. The schema of the objects table is shown in the Fig. 14.

---

[14] https://sqlite.org/index.html.
[15] https://www.sqlalchemy.org/.

```
                          Objects
  id INTEGER NOT NULL

  x REAL, x_err REAL, y REAL, y_err REAL,

  ra TEXT ra_err REAL, dec_TEXT, dec_err REAL,

  ra_deg REAL, dec_deg_REAL,

  inst_mag REAL, inst_mag_err REAL, mag REAL, mag_err REAL,

  ul REAL, refstars TEXT, snr REAL, fwhm_x REAL,

  fwhm_x_err REAL, fwhm_y REAL, fwhm_y_err REAL, flags TEXT,

  catalog TEXT, cat_id TEXT, cat_ra TEXT, cat_ra_err REAL,

  cat_dec TEXT, cat_dec_err REAL,

  cat_ra_deg REAL, cat_dec_deg REAL,

  cat_R1 REAL, cat_B1 REAL, cat_R2 REAL, cat_B2 REAL,

  cat_I REAL, cat_epoch REAL,

  PRIMARY KEY (id)
```

Fig. 14. Schema design of the image database.

For all the objects in the local catalog, region files are created to display the objects in the GUI of the SAOImage DS9[16] software. This feature is implemented by us as the apex.region_generator submodule. A region file contains a row for each object containing its centroid position in the plate or celestial coordinates, depending on the success of the astrometric reduction. Each object displayed as the circular marker labeled by unique id from the database. In future versions of the pipeline, we are planning to migrate to the astropy_regions[17] library, because it has more features.

3.9   Identification of Transients

The object identification is performed with identify_all function from apex.identification submodule which tries to identify the objects by catalog id, brightness or coordinates, i.e. implements the nearest neighbor algorithm. To boost the coordinate querying for nearest objects, we use cKDTree from scipy.spatial submodule, which has a multithread support. The identify_all is executed after the astrometric reduction is done. All unidentified objects are considered as potential candidate transients. To decrease the number of false-positive identifications, we put a constraint on the candidate FWHM: it has not to be significantly different from the point-source average FWHM in the given image. Typically, we observe the fields of afterglow of the discovered GRBs, which coordinates are provided by space gamma and X-ray telescopes, so we want to identify candidates inside a provided circular localization region or countours. For instance, the Swift observatory provides the coordinates of the localization center and its radius

---

[16] https://github.com/SAOImageDS9/SAOImageDS9.
[17] https://astropy-regions.readthedocs.io/en/stable/index.html.

in the GCN. For the purpose of the detection of the Swift transients, we construct a SQLite database, containing trigger date and time, celestial coordinates and error circle radius. Further, we mark all unidentified objects inside a transient localization circle as candidates to that transient. Other space telescopes, say the Fermi GBM, provides transient localization in the HEALPIX format. A HealPix [80] map is the Hierarchical Equal Area isoLatitude Pixelization of the data on sphere stored in FITS files. Each *pixel* contains a probability (including both systematic and statistic uncertainty) of an object located in this pixel to be the transient. The search in the localization contours, provided by the Fermi GBM, is implemented with the `gbm`[18] and `healpy`[19] packages. Initially, `gbm.data.GbmHealPix` is used to read the Fermi HealPix maps. We retrieve the transient localization contours, tracing the 90 per cent confidence level as list of celestial coordinates with `gbm.data.GbmHealPix.confidence_region_path` function. To match the object with those pixels, the `healpy.pixelfunc` routines are applied. The implementation is raw at the moment of writing and in fact is not used. The methods described above make it possible to identify transient candidates based on cross-matching. Another method to find a transient, which is currently not implemented in the pipeline, is the subtraction of the *science* image of the field taken in the current epoch from the *template* image of the same field but taken earlier. The objects appeared in the residual image are transient candidates. Such technique, for example, is used by ZTF project (with ZOGY [84]) and also available in many implementation of *Optimal Image Substraction (OIS)* algorithm [81], for example, in theHOTPANTS[20] [59] tool. We have the raw implementation of the script that runs the HOTPANTS with parameters based on the pipeline created object catalogs for science and reference images, but we do not use in our work.

3.10 Catalog of Transients

We generate text reports on identified candidate transients, which is a convenient method of the distribution of necessary information in a human-readable format. Each report has a conventional name based on the object celestial coordinates in degrees with 5-digit precision. The content of each report includes the following fields:

- UT time of the observation in ISO format at the middle of an exposure;
- Name of the observatory and telescope;
- Single image exposure time and stack size;
- Apparent magnitude and its error;
- Apparent 3σ upper limit at the candidate position;
- Seeing of the image in the arcseconds;
- Photometric reference stars per candidate.

A report file is saved in the same directory with an image file processed.

---

[18] https://fermi.gsfc.nasa.gov/ssc/data/analysis/rmfit/gbm_data_tools/gdt-docs/index.html.
[19] https://github.com/healpy/healpy.
[20] https://github.com/acbecker/hotpants.

3.11   Performance and Accuracy

Performance Estimation. Although, the pipeline usage is not limited by the processing the images from a few telescopes, we present performance profiling only for the two of many (see Tables 1 and 2, values in the tables are rounded up to whole numbers) equipped with wide-field cameras to demonstrate the worst computational scenario. Besides that, estimated time is mainly depends on the number of objects in the field, seeing conditions and the plate scale of a particular CCD.

The testing platform is based on the AMD Ryzen 3500u CPU with 8 threads and 6.7 GB of DDR4-2400 MHz RAM in the dual channel. The disk input-output system is the Western Digital SN 730 SSD with about 512 GB of NVMe memory. The pipeline is planned to be used (and still used) on personal computers rather than on clusters.

Astrometric Accuracy. We obtained a sample of about 10,500 objects detected in the image that presented in both USNO-B1.0 catalog and GAIA DR2 to compare the derived positions of the objects. It is illustrated in the Fig. 15 that astrometric accuracy with GAIA DR2 is 30% better, than with USNO-B1.0, especially in range 18–22 magnitude.

Table 1. Elapsed time of the processing routines measured for AS-32 with FOV = 0.49 degrees^2. There are ≈20000 detected objects in the field at threshold 3σ above the background level.

Routine	Elapsed time[s]
Objects extraction	1
Deblending	100
Isophotal Analysis	5.0
Multi-threaded computation of apertures	70
Multi-threaded PSF fitting of blended groups	80
Multi-threaded PSF fitting of isolated objects	200
Astrometric reduction	50
Photometric reduction	160
Positional look-up for unidentified objects	5
Local catalog creation	10
Identification of transients	10
Creation of local catalog of transients	10
Total time	700

Table 2. Elapsed time of the processing routines for Santel-400 with FOV = 1.82 degrees². The number of detected objects in the field ≈12000.

Routine	Elapsed time[s]
Objects extraction	1
Deblending	60
Isophotal Analysis	4.0
Multi-threaded computation of Apertures	30
Multi-threaded PSF fitting of blended groups	3
Multi-threaded PSF fitting of isolated objects	90
Astrometric reduction	6
Photometric reduction	60
Positional look-up for unidentified objects	2
Local catalog creation	5
Identification of transients	7
Creation of local catalog of transients	8
Total time	310

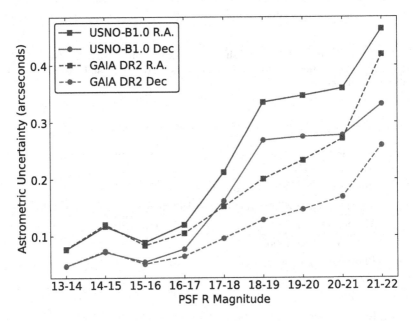

Fig. 15. The astrometric accuracy versus magnitude plot. Blue solid lines are denoting R.A. accuracy and dashed line denote Dec accuracy. Squares and circles are representing USNO-B1.0 and GAIA DR2 points, respectively. (Color figure online)

Higher uncertainties in R.A. axis caused by slightly worse telescope guiding accuracy along it. This is usually a problem with all telescopes. The limiting magnitude at 3 sigma significance level for this particular image was about 21.5–21.8 magnitude, so the centroids of faint objects were computed with a lower precision.

Photometric Accuracy. We obtained a sample of 6400 objects with a photometry tied to the PanSTARRS-PS1 in addition to the USNO-B1.0 sample. Then we compared the photometric accuracy between those two and actually PSF-measured objects. The results of the comparison are shown in the Fig. 16 and Fig. 17.

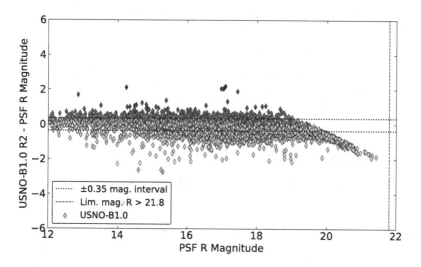

Fig. 16. The photometric accuracy against the USNO-B1.0 stars.

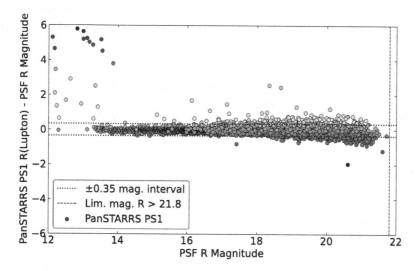

Fig. 17. The photometric accuracy against the PS1 stars.

We note, that for all three samples, only objects that are likely point-like were selected for a proper PSF-photometry. Mean accuracy of PSF-photometry based on up to 4 RP stars is typically about 0.1 magnitude for bright objects and roughly 0.4 magnitudes for objects with SNR = 2.5 − 3.

4   Results and Conclusions

4.1   Status on Implementation of the Pipeline

The following functional units of the pipeline have been implemented (or modified) and tested in the processing of the real survey images that come from IKI GRB Follow-Up Network telescopes:

1. Quality Control and Calibration;
2. Photometric reduction and selection of secondary photometric standards;
3. Objects cataloguing and displaying in GUI of SAOImage DS9;
4. Search and identification of candidates to optical transients;
5. Cataloging of candidates to optical transients.

4.2   Future Plans

The further work is expected to optimize and improve the pipeline, in order to increase accuracy in photometry, candidate OT's identification, data management, and meet the requirement for its automatic operation in real time, being embedded into the telescope software. It is planned to add new local stored and remote catalogs, including local versions of the 2MASS [46] IR catalog, PanSTARRS PS1 and PS2 via interface to the remote database and also SDSS DR9 − DR16 in the same way as PS1/2. A subpixel aperture photometry with masking and weighting will help to improve the photometric performance in cases where PSF-photometry lacks (e.g. measurement of galaxies). Improvement of the OT's identification will be reached with optimal image substraction technique and methods of galaxies's substraction. Machine learning algorithms will help us to classify optical transients from hundreds of thousands of other objects real quick. Modernization of data storage in the form of a distributed database will be required to organize the generated bulky data and distribute them. All of these is a big step in the preparation to the 2022 O4 cycle of LIGO/Virgo/KAGRA observations of BNS mergers.

References

1. Klebesadel, R.W., et al.: Observations of gamma-ray bursts of cosmic origin. ApJ **182**, 85–88 (1973)
2. Hawking, S.W., Israel, W.: Three Hundred Years of Gravitation. Cambridge University Press, Cambridge (1989)
3. Wheaton, W.A., et al.: The direction and spectral variability of a cosmic gamma-ray burst. ApJ **185**, 57–61 (1973). https://doi.org/10.1086/181320

4. Akerlof, C., et al.: Prompt optical observations of gamma-ray bursts. ApJ **532**, 25–28 (2000). https://doi.org/10.1086/312567
5. Kouveliotou, C., et al.: Identification of two classes of gamma-ray bursts. ApJ **413**, 101–104 (1993)
6. Wang, L., Wheeler, J.C.: The supernova-gamma-ray burst connection. ApJL **504**, L87–L90 (1998). https://doi.org/10.1086/311580
7. Kulkarni, S.R., et al.: Radio emission from the unusual supernova 1998bw and its association with the γ-ray burst of 25 April 1998. Nature **395**, 663–669 (1998). https://doi.org/10.1038/27139
8. Lipunov, V., et al.: Master robotic net. Adv. Astron. **2010**, 1(6) (2010). https://doi.org/10.1155/2010/349171
9. Bellm, E.C., et al.: The Zwicky transient facility: system overview, performance, and first results. PASP **131**, 018002 (2019). https://doi.org/10.1088/1538-3873/aaecbe
10. Masci, F.J., et al.: The Zwicky transient facility: data processing, products, and archive. PASP. **131**, 995, 018003 (2018). https://doi.org/10.1088/1538-3873/aae8ac
11. Masci, F.J., et al.: The IPAC image subtraction and discovery pipeline for the intermediate palomar transient factory. Publ. Astron. Soc. Pac. **129**, 014002 (2017). https://doi.org/10.1088/1538-3873/129/971/014002
12. Bertin, E., et al.: The TERAPIX pipeline. **281**, 228 (2002)
13. Bertin, E.: Automatic astrometric and photometric calibration with SCAMP. In: Astronomical Data Analysis Software and Systems XV, vol. 351, p. 112 (2006)
14. Castro-Tirado, A.J., et al.: The burst observer and optical transient exploring system (BOOTES). AAS **138**(3), 583–585 (1999). https://doi.org/10.1051/aas:1999362
15. Dyer, M.J., et al.: A telescope control and scheduling system for the Gravitational-wave Optical Transient Observer (GOTO). In: Observatory Operations: Strategies, Processes, and Systems VII, p. 107040C. International Society for Optics and Photonics (2018). https://doi.org/10.1117/12.2311865
16. Devyatkin, A.V., et al.: Apex I and Apex II software packages for the reduction of astronomical CCD observations. Sol. Syst. Res. **44**, 68–80 (2010). https://doi.org/10.1134/S0038094610010090
17. Kouprianov, V.: Apex II + FORTE: data acquisition software for space surveillance. **39**, 974 (2012)
18. van Dokkum, P.G.: Cosmic-ray rejection by Laplacian edge detection. PASP **113**, 1420–1427 (2001). https://doi.org/10.1086/323894
19. He, Y., et al.: Scan-flood fill(SCAFF): an efficient automatic precise region filling algorithm for complicated regions (2019)
20. Beard, S.M., et al.: The cosmos system for crowded-field analysis of digitized photographic plate scans. MNRAS **247**, 311 (1990)
21. Bertin, E., Arnouts, S.: SExtractor: software for source extraction. Astron. Astrophys. Suppl. **117**, 393–404 (1996). https://doi.org/10.1051/aas:1996164
22. Valdes, F.G., et al.: FOCAS automatic catalog matching algorithms. PASP **107**, 1119 (1995). https://doi.org/10.1086/133667
23. Groth, E.J.: A pattern-matching algorithm for two-dimensional coordinate lists. AsJ **91**, 1244–1248 (1986). https://doi.org/10.1086/114099
24. Monet, D.G., et al.: The USNO-B catalog. Astron. J. **125**, 984–993 (2003). https://doi.org/10.1086/345888
25. Gunn, J.E., et al.: The 2.5 m telescope of the Sloan digital sky survey. ApJ **131**, 2332–2359 (2006). https://doi.org/10.1086/500975

26. Chambers, K.C., et al.: The Pan-STARRS1 surveys. arXiv e-prints. 1612, arXiv:1612.05560 (2016)

27. Lang, D., et al.: Astrometry.net: blind astrometric calibration of arbitrary astronomical images. AJ **139**(5), 1782–1800 (2010). https://doi.org/10.1088/0004-6256/139/5/1782

28. Krimm, H.A., et al.: Swift burst alert telescope data products and analysis software. In: AIP Conference Proceedings, vol. 727, pp. 659–662 (2004). https://doi.org/10.1063/1.1810929

29. Case, G.L., et al.: Monitoring the low-energy gamma-ray sky using earth occultation with GLAST GBM. In: AIP Conference Proceedings, vol. 921, pp. 538–539 (2007). https://doi.org/10.1063/1.2757440

30. Marshall, S., et al.: The ROTSE project. Bull. Am. Astron. Soc. **1290** (1997)

31. Coulter, D.A., et al.: Swope supernova survey 2017a (SSS17a), the optical counterpart to a gravitational wave source. Science **358**, 1556–1558 (2017). https://doi.org/10.1126/science.aap9811

32. Costa, E., et al.: Discovery of an X-ray afterglow associated with the γ-ray burst of 28 February 1997. Nature **387**, 783–785 (1997). https://doi.org/10.1038/42885

33. Krimm, H.A., et al.: GRB 090423: swift detection of a burst. GRB Coordinates Netw. **9198**, 1 (2009)

34. Tanvir, N., et al.: GRB 090423: VLT/ISAAC spectroscopy. GRB Coordinates Netw. **9219**, 1 (2009)

35. Ukwatta, T.N., et al.: GRB 090429B: swift detection of a burst. GRB Coordinates Netw. **9281**, 1 (2009)

36. Cucchiara, A., et al.: A photometric redshift of z 9.4 for GRB 090429B. Astrophys. J. **736**(7) (2011). https://doi.org/10.1088/0004-637X/736/1/7

37. Akerlof, C.W., et al.: The ROTSE-III robotic telescope system. Publ. Astron. Soc. Pac. **115**, 132–140 (2003). https://doi.org/10.1086/345490

38. Akerlof, C.W., McKay, T.A.: GRB990123, early optical counterpart detection. GRB Coordinates Netw. **205**, 1 (1999)

39. Piro, L.: GRB990123, BeppoSAX WFC detection and NFI planned follow-up. GRB Coordinates Netw. **199**, 1 (1999)

40. Collaboration, G., et al.: Gaia data release 2. Summary of the contents and survey properties. Astron. Astrophys. **616**, A1 (2018). https://doi.org/10.1051/0004-6361/201833051

41. Stetson, P.B.: DAOPHOT: a computer program for crowded-field stellar photometry. Publ. Astron. Soc. Pac. **99**, 191 (1987). https://doi.org/10.1086/131977

42. Gorbovskoy, E.: GRB prompt optical observations by Master and Lomonosov. **40**, E1.17-13-14 (2014)

43. Gorbovskoy, E.S., et al.: Prompt, early and afterglow optical observations of five γ-ray bursts: GRB 100901A, GRB 100902A, GRB 100905A, GRB 100906A and GRB 101020A. Mon. Not. R. Astron. Soc. **421**(3), 1874–1890 (2012). https://doi.org/10.1111/j.1365-2966.2012.20195.x

44. Abbott, B.P., et al.: Gravitational waves and gamma-rays from a binary neutron star merger: GW170817 and GRB 170817A. Astrophys. J. Lett. **848**, L13 (2017). https://doi.org/10.3847/2041-8213/aa920c

45. Shapiro, I.I.: Fourth test of general relativity. Phys. Rev. Lett. **13**, 789–791 (1964). https://doi.org/10.1103/PhysRevLett.13.789

46. Skrutskie, M.F., et al.: The two micron all sky survey (2MASS). AJ **131**(2), 1163 (2006). https://doi.org/10.1086/498708

47. Gehrels, N., et al.: The swift gamma-ray burst mission. Astrophys. J. **611**, 1005–1020 (2004). https://doi.org/10.1086/422091

48. Barthelmy, S.D.: Burst Alert Telescope (BAT) on the swift MIDEX mission. In: X-Ray and Gamma-Ray Instrumentation for Astronomy XIII, pp. 175–189. International Society for Optics and Photonics (2004). https://doi.org/10.1117/12.506779
49. Roming, P.W.A., et al.: The swift ultra-violet/optical telescope. In: X-Ray and Gamma-Ray Instrumentation for Astronomy XIII, pp. 262–276. International Society for Optics and Photonics (2004). https://doi.org/10.1117/12.504554
50. Burrows, D.N., et al.: The swift X-ray telescope. Space Sci. Rev. **120**, 165–195 (2005). https://doi.org/10.1007/s11214-005-5097-2
51. Sari, R., et al.: Jets in GRBs. Astrophys. J. **519**(1), L17–L20 (1999). https://doi.org/10.1086/312109
52. Mészáros, P., Rees, M.J.: Optical and long-wavelength afterglow from gamma-ray bursts. Astrophys. J. **476**, 232–237 (1997). https://doi.org/10.1086/303625
53. Shappee, B., et al.: The all sky automated survey for supernovae (ASsAS-SiN). Am. Astron. Soc. Meeting **220**, 432.03 (2012)
54. Bisnovatyi-Kogan, G.S., et al.: Pulsed gamma-ray emission from neutron and collapsing stars and supernovae. Astrophys. Space Sci. **35**(1), 23–41 (1975). https://doi.org/10.1007/BF00644821
55. Colgate, S.A.: Prompt gamma rays and X rays from supernovae. Can. J. Phys. **46**(10), S476–S480 (1968). https://doi.org/10.1139/p68-274
56. Amati, L.: The correlation between peak energy and isotropic radiated energy in GRBs. Il Nuovo Cimento C **28**(3), 251–258 (2005). https://doi.org/10.1393/ncc/i2005-10034-4
57. Minaev, P.Y., Pozanenko, A.S.: The Ep, I-Eiso correlation: type I gamma-ray bursts and the new classification method. Mon. Not. R. Astron. Soc. **492**, 1919–1936 (2020). https://doi.org/10.1093/mnras/stz3611
58. Kochanek, C.S., Piran, T.: Gravitational waves and gamma-ray bursts. Astrophys. J. **417**, L17 (1993). https://doi.org/10.1086/187083
59. Becker, A.: HOTPANTS: high order transform of PSF and template subtraction. Astrophysics Source Code Library. ascl:1504.004 (2015)
60. Andreoni, I., et al.: Fast-transient searches in real time with ZTFReST: identification of three optically-discovered gamma-ray burst afterglows and new constraints on the Kilonova rate. arXiv:2104.06352 (2021)
61. Yao, Y., et al.: ZTF and LT observations of ZTF21aayokph (AT2021lfa), a fast-fading red transient. GRB Coordinates Netw. **29938**, 1 (2021)
62. Andreoni, I., et al.: ZTF21aahifke/AT2021clk: ZTF discovery of an optical fast transient (possible afterglow). GRB Coordinates Network, Circular Service, No. 29446. 9446 (2021)
63. The Physics of Gamma-Ray Bursts - Tsvi Piran. https://ned.ipac.caltech.edu/level5/March04/Piran/Piran7_11.html. Accessed 06 Aug 2021
64. Piran, T.: Gamma-ray bursts and neutron star mergers - possibly the strongest explosions in the universe. In: AIP Conference on Proceedings, vol. 272, pp. 1626–1633 (1992). https://doi.org/10.1063/1.43418
65. Narayan, R., et al.: Gamma-ray bursts as the death throes of massive binary stars. ApJ **395**, L83 (1992). https://doi.org/10.1086/186493
66. Abramovici, A., et al.: LIGO: the laser interferometer gravitational-wave observatory. Science **256**, 325–333 (1992). https://doi.org/10.1126/science.256.5055.325
67. Bradaschia, C., et al.: The VIRGO project: a wide band antenna for gravitational wave detection. Nucl. Inst. Methods Phys. Res. A **289**, 518–525 (1990). https://doi.org/10.1016/0168-9002(90)91525-G
68. Blinnikov, S.I., et al.: Exploding neutron stars in close binaries. Sov. Astr. Let. **10**, 177–179 (1984)

69. LIGO Scientific Collaboration, et al.: Advanced LIGO. Class. Quantum Gravity **32**, 074001 (2015). https://doi.org/10.1088/0264-9381/32/7/074001

70. Acernese, F., et al.: Advanced Virgo: a second-generation interferometric gravitational wave detector. Class. Quantum Gravity **32**, 024001 (2015). https://doi.org/10.1088/0264-9381/32/2/024001

71. Savchenko, V., et al.: INTEGRAL detection of the first prompt gamma-ray signal coincident with the gravitational-wave event GW170817. ApJL **848**, L15 (2017). https://doi.org/10.3847/2041-8213/aa8f94

72. Goldstein, A., et al.: An ordinary short gamma-ray burst with extraordinary implications: fermi-GBM detection of GRB 170817A. ApJ **848**, L14 (2017). https://doi.org/10.3847/2041-8213/aa8f41

73. Hjorth, J., et al.: The distance to NGC 4993: the host galaxy of the gravitational-wave event GW170817. ApJ **848**(2), L31 (2017). https://doi.org/10.3847/2041-8213/aa9110

74. Abbott, B.P., et al.: Gravitational waves and gamma-rays from a binary neutron star merger: GW170817 and GRB 170817A. ApJL **848**, L13 (2017). https://doi.org/10.3847/2041-8213/aa920c

75. Duev, D.A., et al.: Real-bogus classification for the Zwicky Transient Facility using deep learning. MNRAS **489**(3), 3582–3590 (2019). https://doi.org/10.1093/mnras/stz2357

76. Abbott, B.P., et al.: Search for gravitational-wave signals associated with gamma-ray bursts during the second observing run of advanced LIGO and advanced Virgo. ApJ **886**, 75 (2019). https://doi.org/10.3847/1538-4357/ab4b48

77. Flaugher, B., et al.: The dark energy camera. Astron. J. **150**, 150 (2015). https://doi.org/10.1088/0004-6256/150/5/150

78. Bloemen, S., et al.: The BlackGEM array: searching for gravitational wave source counterparts to study ultra-compact binaries. Astron. Soc. Pac. Conf. Ser. **496**, 254 (2015)

79. Pozanenko, A.S., et al.: GRB 170817A associated with GW170817: multi-frequency observations and modeling of prompt gamma-ray emission. Astrophys. J. Lett. **852**, L30 (2018). https://doi.org/10.3847/2041-8213/aaa2f6

80. Gorski, K.M., et al.: HEALPix: a framework for high-resolution discretization and fast analysis of data distributed on the sphere. ApJ **622**(2), 759–771 (2005). https://doi.org/10.1086/427976

81. Alard, C.: Image subtraction with non-constant kernel solutions. astro-ph/9903111 (1999)

82. Laher, R.R., et al.: IPAC image processing and data archiving for the palomar transient factory. Publ. Astron. Soc. Pac. **126**, 674 (2014). https://doi.org/10.1086/677351

83. Lipunov, V.M., et al.: The optical identification of events with poorly defined locations: the case of the Fermi GBM GRB 140801A. Mon. Not. R. Astron. Soc. **455**(1), 712–724 (2016). https://doi.org/10.1093/mnras/stv2228

84. Zackay, B., et al.: Proper image subtraction-optimal transient detection, photometry, and hypothesis testing. ApJ **830**(1), 27 (2016). https://doi.org/10.3847/0004-637X/830/1/27

VALD in Astrophysics

Yury Pakhomov$^{(\boxtimes)}$ (iD) and Tatiana Ryabchikova$^{(\boxtimes)}$ (iD)

Institute of Astronomy, Russian Academy of Sciences, 119017 Moscow, Russia
{pakhomov,ryabchik}@inasan.ru

Abstract. Vienna Atomic Line Database (VALD) is a most popular among astrophysicists for studying stars, stellar systems, interstellar medium, its chemical composition, evolution and kinematics. The article briefly describes the evolution and modern state of the VALD. The database contains parameters for millions of atomic and molecular lines which provide possibility of synthetic spectra calculation. VALD is an active member of international project "Virtual Atomic and Molecular Data Centre" (VAMDC).

Keywords: Database · Atomic data · Molecular data · Stellar spectra · Spectral lines · Hyperfine splitting · Interstellar medium · Astronomical tools · Hollow-cathode lamp · VAMDC

1 VALD Database

Vienna Atomic Line Database (VALD) is a database of atomic parameters of spectral lines. The main fraction of radiation in the optical range comes from stars and stellar systems (stellar clusters, galaxies). Spectroscopy is the most powerful tool to explore the Universe because more than 90% of information we received from spectra, therefore, we need a huge amount of atomic and molecular data as well as powerful tools to model the spectra of cosmic bodies, mostly, stars, but also interstellar medium. Modelling the spectra of stars allows us to study the chemical composition of stars and galaxies, their dynamics and evolution. However, for adequate modelling, it is necessary to have the parameters of the cosmic material and atomic parameters of hundreds of millions of spectral lines that produce the observed spectra. To assess the quality of atomic parameters and facilitate access to them, special databases have been created. One of the most famous is the Atomic Spectra Database [6], created at the National Institute of Standards and Technology (NIST) in the USA. NIST ASD contains mainly data obtained from laboratory experiments, which are very important in terms of accuracy, but, unfortunately, provide only a small part of the data required for detailed spectral analysis.

In 1991, in Vienna (Austria), a group of astrophysicists from Austria, Russia and Sweden began to create an open access database of atomic parameters of

A. Pozanenko et al. (Eds.): DAMDID/RCDL 2021, CCIS 1620, pp. 135–148, 2022.
https://doi.org/10.1007/978-3-031-12285-9_8

spectral lines to facilitate stellar spectra calculations. The Vienna Atomic Line Database (VALD) quickly became one of the most popular database among astrophysicists because it contains experimental atomic and molecular data for detailed analysis of stellar spectra, as well as a huge amount of less accurate theoretical data for opacity calculations. VALD compiles data from various sources, including NIST ASD, lists of theoretical calculations by Kurucz [8] and many individual articles. The first version of VALD [16] became available to users in 1994. It contained data for about one million spectral lines. In 1999 the second version of the database [7] started to operate. The number of lines were doubled (about two million transitions between levels with experimentally measured energies), and in addition, theoretical calculations for more than 40 million predicted lines were added. In 2012, the third version of VALD [18] was prepared, containing about 250 million atomic lines and about 1.5 billion lines of few diatomic molecules (TiO, CN, CH, C_2, O_2, SiH, FeH, MgH, etc.) and H_2O. The database is permanently updated. In 2016 the isotopic splitting was included in the VALD, and later a separate database of hyperfine splitting coefficients was created and joined to VALD [14] (Fig. 1).

To use the VALD a registration of user by email is required. Each user can be registered under several email addresses which will be used as return address for queries on the website.

Tools. VALD provides special extraction tools available for registered users:

- *Showline* - all information about selected spectral line from available sources and atomic data recommended for use.
- *Extract all* - the full linelist in the selected wavelength range.
- *Extract element* - then same, only for selected element or ion.
- *Extract stellar* - the list of observable spectral lines for star with given atmospheric parameters (Fig. 2).

These tools are implemented as corresponded request by email. User can independently form a request according to the template. However, it is easier to do this through the form on the website. The short answer with requested data will be sent via email, while the long as FTP link.

All requests are executed asynchronously. There is a queue of jobs that are processed sequentially. The *Showline* request also available as online tool on the website (Fig. 3). In this case (this request is the simplest), the information immediately output to the website.

The VALD output linelist serves as an input file for stellar spectrum synthesis realized in different codes.

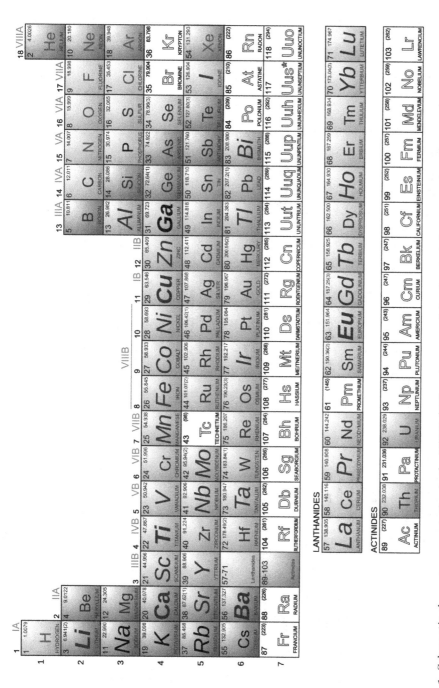

Fig. 1. Schematic element representation in the VALD (different groups of the chemical elements are marked with a colored background). Items with white background are missing in the database. Elements for which data on stable isotopes are available are marked by bold font, elements with data on hyperfine splitting are marked by italic font. (Color figure online)

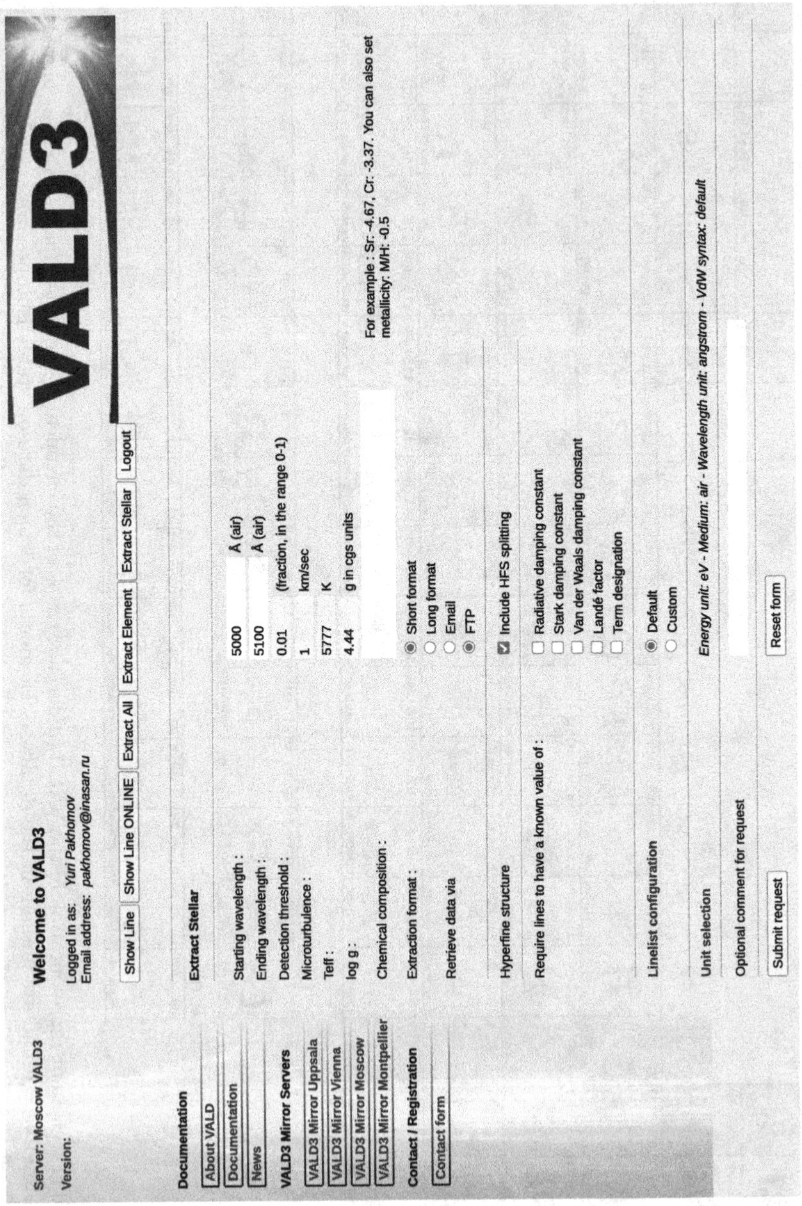

Fig. 2. Web interface for the request of *Select stellar*.

Modern State. To date, VALD contains precise laboratory wavelengths for 1.804 million spectral lines of 299 atoms and ions of 80 elements (Fig. 1). The transition probabilities (oscillator strengths) for part of these lines were determined in laboratory experiments, and for the rest they were calculated [8]. The bibliography contains 1384 sources. These data are used for detailed spectral analysis. Theoretical calculations for 254 million predicted atomic lines are used in opacity calculations for modelling stellar atmospheres. In addition, VALD contains 1.5 billion observed and predicted molecular lines. Almost all elements are presented in several ionization states (up to IX for iron group).

The wavelength range from 10 Å to 1 mm covers a fairly large part of the electromagnetic radiation: from X-ray to near radio.

For most elements, the spectral line parameters (primarily the wavelength and oscillator strength) are given for the centre of gravity of the mixture of isotopes. For six elements Li, Ca, Ti, Cu, Ba, Eu (shown in bold in Fig. 1) atomic parameters for lines of stable isotopes are available.

The atomic levels of isotopes with an odd mass number are subject to the hyperfine splitting effect (hfs). As a result, the transition between such levels is divided into a number of components, which should be taken into account when calculating the synthetic spectrum. The hyperfine splitting coefficients A and B are collected in a separate SQL database for 29 isotopes ^{6,7}LiI, ^{23}NaI, ^{27}AlI-II, ^{39,40,41}KI, ^{45}ScI-II, ^{47,49}TiI-II, ^{50,51}VI, ^{51}VII, ^{55}MnI-II, ^{57}FeI, ^{59}CoI-II, ^{61}NiI, ^{63,65}CuI, ^{67}ZnI-II, ^{85,87}RbI, ^{89}YII, ^{135,137}BaII, ^{139}LaII, ^{151,153}EuII, and ^{159}TbII; in total for >2000 levels. These elements are shown by italics characters in Fig. 1. If splitting coefficients are available for both levels of the transition, the calculation of individual hfs components is carried out "on the fly" by the user request.

VALD has four main mirror-sites located at

- Uppsala (Sweden) http://vald.astro.uu.se/~vald/php/vald.php
- Moscow http://vald.inasan.ru/~vald3/php/vald.php
- Vienna (Austria) http://vald.astro.univie.ac.at/~vald3/php/vald.php
- Montpellier (France) http://vald.oreme.org/~vald/php/vald.php

These sites are daily synchronized. In addition, there are local VALD copies in Germany and in the United States. VALD has a detailed documentation for users and also for the installation of new mirror.

Users. VALD has about 2800 registered users representing more than a hundred scientific institutions from 74 countries of all continents, excluding Antarctica (Fig. 4). Most of the users are scientists and students in the field of astrophysics, plasma physics, but there are also astronomy amateurs. All of them make over 1,500 requests every month. The VALD popularity arises from the wide range of requests for data retrieval, which is absent in other similar collections. The most popular is the 'Extract Stellar' query, which return to the user atomic parameters for those lines that make a real contribution to the spectrum of a star with given atmospheric parameters (effective temperature, gravity, and metallicity). The threshold is set by the user. This type of request can significantly reduce the amount of data received by the user.

These are the spectral lines found (VALD ranking is shown in parentheses below each entry):

Database reference for the line	Wavelength [Å]	Element Ion	log gf	Elow [cm^-1]	Jlow	Eup [cm^-1]	Jup	Lande eff	Lande low	Lande up	γ Rad	γ Stark	γ VdW	Lower level coupling electronic configuration Term designation	Upper level coupling electronic configuration Term designation	Source Accuracy Comment
Li 1 - K 5 Bell	6707.7635 (2)	Li 1	-0.002 (2)	0.0000 (1)	0.5	14903.9835 (2)	1.5	99.000 (2)	99.00	99.00	7.560 (2)	-5.780 (2)	0.000 (2)	LS 1s2.2s 2S	LS 1s2.2p 2P*	Li 1 - K 5 Bell (2) (7)Li
Van der Waals data	6707.7635 (0)	Li 1	0.000 (1)	0.0000 (1)	0.5	14903.9835 (1)		99.000 (1)	99.00	99.00	0.000 (1)	0.000 (1)	346.236 (9)			Van der Waals da (1) Li
Li 1 - K 5 Bell	6707.9145 (2)	Li 1	-0.303 (2)	0.0000 (1)	0.5	14903.6481 (2)	0.5	99.000 (2)	99.00	99.00	7.560 (2)	-5.780 (2)	0.000 (2)	LS 1s2.2s 2S	LS 1s2.2p 2P*	Li 1 - K 5 Bell (2) (7)Li
Van der Waals data	6707.9145 (0)	Li 1	0.000 (1)	0.0000 (1)	0.5	14903.6481 (1)		99.000 (1)	99.00	99.00	0.000 (1)	0.000 (1)	346.236 (9)			Van der Waals da (1) Li
Li 1 - K 5 Bell	6707.9215 (2)	Li 1	-0.002 (2)	0.0000 (1)	0.5	14903.6321 (2)	1.5	99.000 (2)	99.00	99.00	7.560 (2)	-5.780 (2)	0.000 (2)	LS 1s2.2s 2S	LS 1s2.2p 2P*	Li 1 - K 5 Bell (2) (6)Li
Van der Waals data	6707.9215 (0)	Li 1	0.000 (1)	0.0000 (1)	0.5	14903.6321 (1)		99.000 (1)	99.00	99.00	0.000 (1)	0.000 (1)	346.236 (9)			Van der Waals da (1) Li
Li 1 - K 5 Bell	6708.0725 (2)	Li 1	-0.303 (2)	0.0000 (1)	0.5	14903.2968 (2)	0.5	99.000 (2)	99.00	99.00	7.560 (2)	-5.780 (2)	0.000 (2)	LS 1s2.2s 2S	LS 1s2.2p 2P*	Li 1 - K 5 Bell (2) (6)Li
Van der Waals data	6708.0725 (0)	Li 1	0.000 (1)	0.0000 (1)	0.5	14903.2968 (1)		99.000 (1)	99.00	99.00	0.000 (1)	0.000 (1)	346.236 (9)			Van der Waals da (1) Li

The data above should be combined to the following set of lines (The output in long format skipping Lande factors is displayed).
Numbers in front of the source keys refer to the line list names below.

Wavelength [Å, air]	Element Ion	log gf	Elow [cm^-1]	Jlow	Eup [cm^-1]	Jup	Lande	γ Rad	γ Stark	γ VdW	Lower level coupling electronic configuration Term designation	Upper level coupling electronic configuration Term designation	Reference for wavelength
6707.7635 1 iso:REB	Li 1	-0.036 1 gf:YD	0.0000 1 YD	0.5	14903.9835 1 YD	1.5	99.00 1 YD	7.560 1 YD	-5.780 1 YD	0.000 1 YD	LS 1s2.2s 2S 1 YD	LS 1s2.2p 2P* 1 YD	Li 1 - K 5 Bell (7)Li
6707.9145 1 iso:REB	Li 1	-0.337 1 gf:YD	0.0000 1 YD	0.5	14903.6481 1 YD	0.5	99.00 1 YD	7.560 1 YD	-5.780 1 YD	0.000 1 YD	LS 1s2.2s 2S 1 YD	LS 1s2.2p 2P* 1 YD	Li 1 - K 5 Bell (7)Li
6707.9215 1 iso:REB	Li 1	-1.122 1 gf:YD	0.0000 1 YD	0.5	14903.6321 1 YD	1.5	99.00 1 YD	7.560 1 YD	-5.780 1 YD	0.000 1 YD	LS 1s2.2s 2S 1 YD	LS 1s2.2p 2P* 1 YD	Li 1 - K 5 Bell (6)Li
6708.0725 1 iso:REB	Li 1	-1.423 1 gf:YD	0.0000 1 YD	0.5	14903.2968 1 YD	0.5	99.00 1 YD	7.560 1 YD	-5.780 1 YD	0.000 1 YD	LS 1s2.2s 2S 1 YD	LS 1s2.2p 2P* 1 YD	Li 1 - K 5 Bell (6)Li

Key to references:
1 Li 1 - K 5 Bell

Fig. 3. Web page of the *Showline* online tool for Li 1 lines in wavelength range of $\lambda = 6707 \pm 2$ Å. Top table is a list from all available sources. Bottom table is a list merged corresponds to used configuration file.

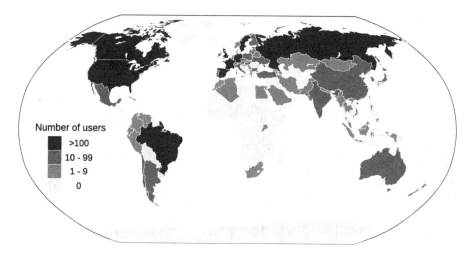

Fig. 4. Map of countries with VALD users.

Demand of VALD. The citation statistics of the main publications reflects the relevance of the database in the scientific community. Currently, VALD is referenced by more than 200 times per year (about 4 times per week). According to ADS (https://ui.adsabs.harvard.edu/) VALD citation index is approaching to 3000.

2   Using the VALD

2.1   Stellar Astrophysics

Originally, VALD was designed for stellar spectrum analysis and opacity calculations in modelling of stellar atmospheres.

To model the stellar spectrum we need the accurate atomic data, model of stellar atmosphere, and abundances of chemical elements. The *Synth* code is used for this purpose [22]. The BinMag program [4] based on VALD line list often used for comparing stellar observed with theoretical spectra. This is a front-end widget for Synth/SynthMag/Synmast/Synth3/SME spectrum synthesis codes. An example of a small (12 Å) region in spectra of three stars with different effective temperatures is shown in Fig. 5. To synthesize this small spectral part for different stars one needs to have accurate atomic and molecular data for more than 500 lines. This illustrates the fact that we need a lot of precise and accurate atomic data.

In high resolution spectroscopy the accurate atomic data obtained in laboratory measurements used to calculate of spectral line profiles while the predicted atomic data used to calculate the continuum spectrum. These opacities can be calculated in two variants. The first is ODF (opacity distribution function) - the cumulative opacities within a small region (about 20 Å in the optical range).

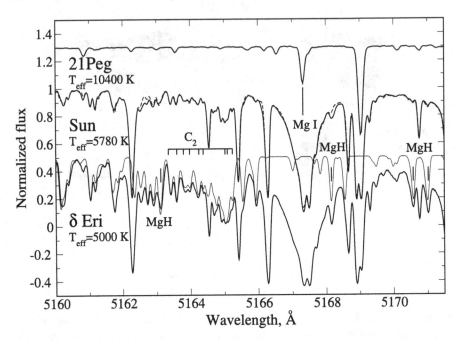

Fig. 5. Example of spectra of three stars with different effective temperatures. Solar synthetic spectrum is shown by grey line. A contribution of molecular lines to the spectrum of δ Eri is indicated by thin line.

Once precalculated ODF for a grid of stellar parameters can be used to construct the models of stellar atmosphere with arbitrary parameters. The program *ATLAS9* is based on ODF [2]. Little time is spent on the calculation of such models. This approach is applicable for spectral analysis of normal stars. The second variant takes into account the individual chemical composition of stellar atmosphere and used in the program *LLmodel* [21]. For each star it is necessary to calculate its own model corresponds to its abundances, so spend much more time. This approach is applicable in case of stars with abnormal chemical composition and stratification of chemical elements in the atmosphere.

Using the precalculated grids of stellar atmospheres and atomic/molecular data from VALD one may determine parameters of stellar atmosphere (the effective temperature, the surface gravity, the turbulent velocities, and also the rotation velocity) and abundances of chemical elements from the observed stellar spectrum. Spectroscopy Made Easy (SME) package [23] based on the VALD lists often used to analyse the stellar spectra. SME can fit the large wavelength regions of observational spectrum to find the solution relative stellar parameters by the minimization procedure. The preparation of spectral mask is an important part to get the correct result. This mask marks the regions in the spectrum which are sensitive to the changes of stellar parameters (Fig. 6). The uncertainty of the derived parameters and chemical element abundances depend on both the quality of the spectral observations and the quality of the atomic parameters [19].

VALD also used for analysis of the complex spectra. In case of the presence of a strong magnetic field, the line profiles will be distorted by Zeeman effect. Least-squares deconvolution (LSD) method is used to recover the unaffected spectrum [5]. The quality of this procedure strongly depends on accuracy of theoretical profiles of many preselected spectral lines and hence on the atomic data accuracy.

The spectra of binary stars where the spectra of both components are seen also needs to be disentangled before analysis [20]. For this, one spectrum is not enough, we must have several observations in different phases of orbital rotation. The radial velocities of the components must be taken into account. After disentangling procedure we can analyse the separated spectra as spectra of a single star.

The results obtained from spectral analysis with VALD are used to study stellar physics [17], chemical evolution of stellar systems including clusters [12], moving groups [11], our Galaxy [24] and other galaxies [9]. Chemical composition also is employed as an indicator of membership in binary systems and open clusters.

2.2   Astrophysics of Interstellar Medium (ISM)

The interstellar spectral absorption lines pollute the observed stellar spectra. However, often it is a single method to trace ISM. The radiation of distant stars is good background for appearance of ISM spectral lines. To carefully extract the ISM spectrum we need to model accurately the spectra of background stars [13]. The typical ISM spectral lines are resonance atomic lines of Na I, Ca I, Ca II, K I, and sometimes molecular lines of CH, CH^+, and more complex compounds such as polyaromatic hydrocarbons which form in the spectra diffusional interstellar bands. The analysis of ISM spectra give us three parameters of interstellar clouds: the radial velocity, the column density, and the turbulent velocity. The distribution of these parameters reflects the physical condition in the observed direction.

2.3   Optical Tools

The VALD data was used to identify lines in spectra of hollow-cathode lamps and create a list of reference lines [10,15]. These lamps are applied for wavelength calibration of observed spectra. However, sometimes the technology of lamp production or lamp marking is malfunctioning. In this case, in spectrum of the lamp unknown lines are appeared. We developed a statistical method to identify unknown lines by some chemical element using only wavelength. The method is based on estimation of fraction of the spectral lines of any elements with decrements of value of detection accuracy. The initial value corresponds to typical accuracy of the current wavelength calibration.

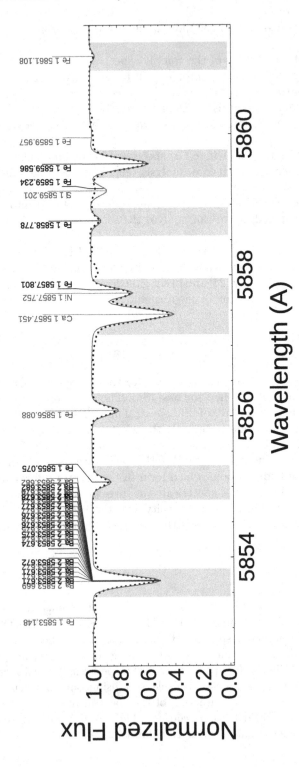

Fig. 6. Example of the part of the mask choice realised in SME program package for automatic determination of the stellar parameters and the element abundances.

We extract all elements from VALD which wavelengths corresponds to wavelengths of unknown lines taking into account the accuracy value. If the value is halved then number of spectral lines of the random element also is halved, while elements with observed lines will have a greater portion. In Fig. 7 shown the element portions for two lamps used in Lick observatory (USA). So, we conclude that the first lamp has titanium cathode and filled by argon, the second lamp has cathode made of two metals (thorium and tantalum) and filled also by argon. To check the correctness, we used the oscillator strengths f and upper energy levels E_{up} from VALD. The intensity of emission lines should be $I \propto \frac{gf}{\lambda^3}e^{(-E_{up}/kT)}$, where g is the statistical weight, λ is the wavelength, and T is the gas temperature in the lamp. The lines of random elements will show the large scatter while the real element will satisfy this dependence and all real elements will have equal slopes (Fig. 8).

Having the identified elements and the lamp temperature we created the list of available spectral lines. The list is needed for accurate wavelength calibration.

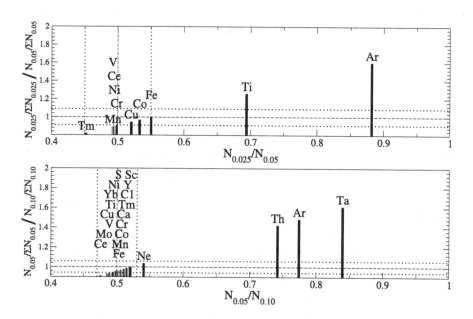

Fig. 7. Relative change in the proportion of detected elements depending on the relative change in their number when the detection area is halved. The dotted line marks the mean values with errors for the elements missing in the calibration lamp. Top: data for LICK-HCL-002 lamp. Bottom: data for the ThAr02 lamp.

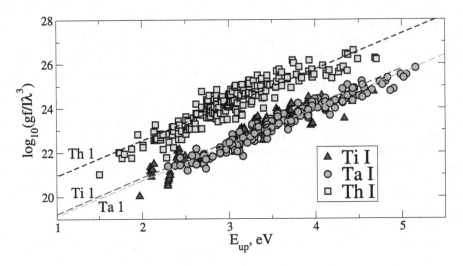

Fig. 8. Checking identified spectral lines of ThAr02 hallow-cathode lamp. Linear regression for each material is marked by dashed line.

3   VALD in VAMDC

VALD developer's team is an active member of international project Virtual Atomic and Molecular Data Centre (VAMDC) (https://portal.vamdc.eu/) [1,3] which incorporates more than 40 atomic and molecular databases. Purpose of the project is simplifying the search for atomic and molecular data. VAMDC provides the unified standard interface between users and different databases (nodes). The query from user is sent to the nodes containing the required data. The responses from these nodes return in a uniform format. VALD database is divided in two VAMDC nodes. The first node is located in Uppsala and provides atomic data for all available spectral lines including the predicted ones used in opacity calculations, while the second node is located in Moscow and provides atomic and molecular data for fine spectrum analysis. In addition to standard requests, a mechanism is being developed for specific requests in VALD (e.g. select stellar) via VAMDC portal.

4   Conclusion

Since its creation, the database has significantly increased in volume and functionality, the number of users is constantly growing. In this article shown the use of the database in astrophysics. At the present time, the VALD finds new applications in different fields of physics due to feedback and regularly updates. The database will continue to be incorporate into international projects.

References

1. Albert, D., Antony, B.K., Ba, Y.A., et al.: A decade with VAMDC: results and ambitions. Atoms **8**(4), 76 (2020). https://doi.org/10.3390/atoms8040076
2. Castelli, F., Kurucz, R.L.: New grids of ATLAS9 model atmospheres. In: Piskunov, N., Weiss, W.W., Gray, D.F. (eds.) Modelling of Stellar Atmospheres, vol. 210, p. A20 (2003)
3. Dubernet, M.L., Antony, B.K., Ba, Y.A., et al.: The virtual atomic and molecular data centre (VAMDC) consortium. J. Phys. B: At. Mol. Phys. **49**(7), 074003 (2016). https://doi.org/10.1088/0953-4075/49/7/074003
4. Kochukhov, O.: BinMag: widget for comparing stellar observed with theoretical spectra (2018)
5. Kochukhov, O., Makaganiuk, V., Piskunov, N.: Least-squares deconvolution of the stellar intensity and polarization spectra. A&A **524**, A5 (2010). https://doi.org/10.1051/0004-6361/201015429
6. Kramida, A., Ralchenko, Y., Reader, J.: NIST ASD Team: NIST Atomic Spectra Database (ver. 5.8) (2019). https://doi.org/10.18434/T4W30F
7. Kupka, F., Piskunov, N., Ryabchikova, T.A., Stempels, H.C., Weiss, W.W.: VALD-2: progress of the Vienna atomic line data base. A&A Sup. **138**, 119–133 (1999). https://doi.org/10.1051/aas:1999267
8. Kurucz, R.L.: Atomic parameters database. http://kurucz.harvard.edu/atoms
9. Mashonkina, L., Jablonka, P., Sitnova, T., Pakhomov, Y., North, P.: The formation of the Milky Way halo and its dwarf satellites; a NLTE-1D abundance analysis. II. Early chemical enrichment. Astron. Astrophys. **608**, A89 (2017). https://doi.org/10.1051/0004-6361/201731582
10. Pakhomov, Y.V.: A list of tantalum lines for the wavelength calibration of the Hamilton echelle spectrograph. Astron. Rep. **59**(10), 952–958 (2015). https://doi.org/10.1134/S1063772915100054
11. Pakhomov, Y.V., Antipova, L.I., Boyarchuk, A.A.: Chemical abundance analysis for the atmospheres of red giants in the Hercules moving group. Astron. Rep. **55**(3), 256–266 (2011). https://doi.org/10.1134/S106377291103005X
12. Pakhomov, Y.V., Antipova, L.I., Boyarchuk, A.A., et al.: A study of red giants in the fields of open clusters. Cluster members. Astron. Rep. **53**(7), 660–674 (2009). https://doi.org/10.1134/S1063772909070087
13. Pakhomov, Y.V., Chugai, N.N., Iyudin, A.F.: Interstellar absorptions and shocked clouds towards the supernova remnant RX J0852.0-4622. MNRAS **424**(4), 3145–3155 (2012). https://doi.org/10.1111/j.1365-2966.2012.21476.x
14. Pakhomov, Y.V., Ryabchikova, T.A., Piskunov, N.E.: Hyperfine splitting in the VALD database of spectral-line parameters. Astron. Rep. **63**(12), 1010–1021 (2019). https://doi.org/10.1134/S1063772919120047
15. Pakhomov, Y.V., Zhao, G.: Wavelength calibration of the Hamilton echelle spectrograph. Astron. J. **146**(4), 97 (2013). https://doi.org/10.1088/0004-6256/146/4/97
16. Piskunov, N.E., Kupka, F., Ryabchikova, T.A., Weiss, W.W., Jeffery, C.S.: VALD: the Vienna atomic line data base. A&A Sup. **112**, 525 (1995)
17. Romanovskaya, A., Ryabchikova, T., Shulyak, D., et al.: Fundamental parameters and evolutionary status of the magnetic chemically peculiar stars HD 188041 (V1291 Aquilae), HD 111133 (EP Virginis), and HD 204411: spectroscopy versus interferometry. MNRAS **488**(2), 2343–2356 (2019). https://doi.org/10.1093/mnras/stz1858

18. Ryabchikova, T., Piskunov, N., Kurucz, R.L., et al.: A major upgrade of the VALD database. Phys. Scr. **90**(5), 054005 (2015). https://doi.org/10.1088/0031-8949/90/5/054005

19. Ryabchikova, T., Piskunov, N., Pakhomov, Y., et al.: Accuracy of atmospheric parameters of FGK dwarfs determined by spectrum fitting. MNRAS **456**(2), 1221–1234 (2016). https://doi.org/10.1093/mnras/stv2725

20. Ryabchikova, T., Zvyagintsev, S., Tkachenko, A., et al.: Fundamental parameters and abundance analysis of the components in the SB2 system HD 60803. MNRAS **509**(1), 202–211 (2022). https://doi.org/10.1093/mnras/stab2891

21. Shulyak, D., Tsymbal, V., Ryabchikova, T., Stütz, C., Weiss, W.W.: Line-by-line opacity stellar model atmospheres. A&A **428**, 993–1000 (2004). https://doi.org/10.1051/0004-6361:20034169

22. Tsymbal, V., Ryabchikova, T., Sitnova, T.: Software for NLTE spectrum fitting. In: Kudryavtsev, D.O., Romanyuk, I.I., Yakunin, I.A. (eds.) Physics of Magnetic Stars. Astronomical Society of the Pacific Conference Series, vol. 518, pp. 247–252 (2019)

23. Valenti, J.A., Piskunov, N.: Spectroscopy made easy: a new tool for fitting observations with synthetic spectra. A&A Sup. **118**, 595–603 (1996)

24. Zhao, G., Mashonkina, L., Yan, H.L., et al.: Systematic non-LTE study of the -.6 ≤ [Fe/H] ≤ 0.2 F and G dwarfs in the solar neighborhood. II. Abundance patterns from Li to Eu. Astrophys. J. **833**(2), 225 (2016). https://doi.org/10.3847/1538-4357/833/2/225

Data Analysis in Material and Earth Sciences

Machine Learning Application to Predict New Inorganic Compounds – Results and Perspectives

Nadezhda Kiselyova[1] ⓘ, Victor Dudarev[1,2(✉)] ⓘ, and Andrey Stolyarenko[1]

[1] A. A. Baikov Institute of Metallurgy and Materials Science of RAS (IMET RAS), Moscow 119334, Russia
kis@imet.ac.ru, vdudarev@hse.ru
[2] HSE University, Moscow 109028, Russia

Abstract. A brief overview of the problems is given in the field of inorganic chemistry and materials science, solved using machine learning (ML). The main ML methods limitations and the subject area peculiarities are considered that must be taken into account when using ML. Solved problems examples of new inorganic compounds design and the results of comparing predictions with new experimental data are given. Systems developed by the authors are considered that aimed at not yet obtained inorganic compounds design, based on ML methods, as well as promising directions for such systems development in order to improve the predictions accuracy for new substances and their corresponding properties values estimations.

Keywords: Machine learning · Databases · Inorganic substances physical properties prediction

1 Introduction

The problem of predicting the new compounds formation and calculating their properties values is one of the most important problems in inorganic chemistry. Any successful attempt to design compounds that have not yet been obtained is of great theoretical and practical importance. The solution to the new inorganic compounds design problem is associated with many difficulties. The main one is the complexity of the dependencies connecting the inorganic compounds properties with chemical elements properties. The traditional way of solving this problem is associated with quantum mechanical methods. As a rule, various approximate methods are used, which very often are not capable to give the desired results for not yet obtained compounds.

It should be remembered that chemistry has accumulated a big data on the inorganic substances' properties. There are complex periodic dependencies between the compound's properties and the properties of the elements that make up them. Moreover, it is obvious that already known compounds must obey these periodic regularities. One of

© Springer Nature Switzerland AG 2022
A. Pozanenko et al. (Eds.): DAMDID/RCDL 2021, CCIS 1620, pp. 151–165, 2022.
https://doi.org/10.1007/978-3-031-12285-9_9

the most promising ways to search for such regularities is the machine learning methods application. They make it possible to find such periodic regularities based on the analysis of already known substances information, collected in databases on the properties of inorganic substances and materials (DB PISM). The found regularities are used to design not yet obtained inorganic compounds – analogs of already existing compounds.

In the mid-sixties, the idea to apply machine learning methods to predict new inorganic compounds was proposed by the colleagues from our Institute of Metallurgy – Savitskii and Gribulya [1]. Already the first calculations gave amazing results: the prediction accuracy was about 90% for not yet obtained binary compounds. In that case, only widely known data were used for prediction, such as data on the electrons' distribution over the energy shells of elements' atoms. We started similar studies in the seventies at the Moscow State University [2]. Problems of predicting more complex ternary compounds were solved. This work was further continued at the Institute of Metallurgy.

2   The Basic Peculiarities of Machine Learning Applications to Prediction Problems Solution for New Inorganic Compounds

Let us first agree on terms and explain some definitions we use.

An *object* is a chemical system (inorganic compound, solid solution, heterogeneous mixture, etc.) formed by *components* (chemical elements or simpler inorganic compounds), represented in the computer memory as a set of *attribute* values (components' properties) and the value of a given *target* physical or chemical parameter (property).

Qualitative (categorical) property – an object's parameter, represented by some discrete value (for example, a crystal structure type or a space symmetry group).

Quantitative property – an object's parameter, represented by a numerical (scalar) scaled variable (for example, the crystal lattice parameter value or the melting point value).

A *training sample* is a matrix, each row contains a set of values for the parameters of the components that make up an object (for example, a compound) for which the target property value is known.

The goal of *machine learning* (ML): based on the analysis of the information contained in the training sample, find regularities that allow to associate the target parameter with a set of feature values and predict the target parameter value for objects that have not yet been studied.

Depending on the target parameter nature, the tasks, solved in the ML process, can be conditionally divided into two types:

- if the target parameter has a discrete nature (a finite set of integers or designations of some precedents classes is a categorical property), then such tasks are related to machine learning (or precedent based pattern recognition) tasks;
- if the target parameter is presented in the real numbers form (quantitative property), then we are dealing with the regression estimation problem.

Other target parameter forms are also possible, for example, images, temporal series, sound recordings, etc.

Like all theoretical methods, machine learning has its limitations. The main one is the need for a sufficient and representative set of objects for machine learning. When predicting qualitative properties, it is necessary to fulfill (although not very strict) the main hypothesis of pattern recognition methods – the compactness hypothesis. In machine learning application to inorganic chemistry, the fulfillment of the latter is a consequence of the Periodic Law. When predicting quantitative properties, it is important to have a statistical relationship between the target parameter and the parameters of the components, which is reflected in the multiple determination coefficient value (R^2).

When developing and applying machine learning methods to inorganic chemistry, one should take into account the peculiarities of the problems being solved and, first of all, the limited volumes of training samples, periodic dependences of the attributes values (chemical elements properties) on the atomic number and, as a consequence, the multicollinearity of the system of elements properties (attributes), the target dependencies nonsmoothness, the attribute description large dimension in combination with a small number of precedents for machine learning, a significant difference in the sizes of objects classes, the gaps presence in some objects attribute description and, as in any experimental science, errors in attributes values and in the target parameter values also.

In order to form a representative training sample, it is necessary to have large databases on the properties of inorganic substances and materials, preferably containing information assessed by experts. As a rule, DB PISM contain "raw" information, and the laborious task of expert assessment of its correctness falls on a specialist who trains artificial intelligence programs. The DB PISM developed by us (http://www.imet-db.ru/) contains information on more than 85 thousand inorganic substances. Databases are integrated with information systems developed in Russia and Japan [3]. An expert assessment of the data correctness is greatly facilitated by the fact that our databases include about 36 thousand pdf-publications from which the data has been extracted. An important feature of the databases developed at IMET is that, in contrast to the existing information systems in the inorganic chemistry and materials science domain, their conceptual structure provides a very fast automated search for initial information for machine learning. For example, it takes 1–2 min to find information about all known oxide spinels in our Phases database. Such a procedure for the well-known ICSD database takes several weeks of painstaking work carried out by the qualified researcher.

The integrated DB PISM subsystem [3] is an information component for the systems for the inorganic compounds computer design developed by us: IAS [4] and ParIS [5]. The first is intended for predicting the inorganic compounds categorical properties, and the second is for quantitative parameters estimation.

3 Some Machine Learning Application Results to Inorganic Chemistry and Materials Science

In recent years, machine learning has been widely used in various fields of inorganic chemistry and materials science, including industrial technologies for the production and processing of inorganic materials. Over the past decade, thousands of articles have been published on the ML use for solving the above problems. One of the accelerators of this process was MGI [6] and similar programs adopted and financed by developed countries [7, 8], one of the main goals of which was the widespread calculation methods introduction. An important role was also played by the free software packages availability for machine learning, for example, [9, 10]. It was shown that the ML methods application allows one to quickly and fairly accurately solve most problems, including those that are poorly solved by other methods.

Based on our experience, we will consider some of the possibilities that machine learning methods provide for solving problems of predicting not yet obtained inorganic compounds and estimating their properties, and also touch on industrial applications.

3.1 Predicting Compound Formation

Usually, the problem to be solved is whether a stable compound made of a certain atoms composition exists under given external conditions (usually at room temperature and atmospheric pressure). To solve such a classical problem of predicting categorical properties, precedent based pattern recognition methods are successfully used. For example, the method proposed by Devingtal' [11] was used to predict new A_3B composition binary compounds [1]. Gradient boosting, random trees ensembles, and other algorithms were applied to predict ternary compounds of different compositions [12].

One of the first tasks, which was solved by us almost half a century ago, was the prediction of yet not obtained ABO_3 composition compounds [2]. One of the most effective algorithms of that time was used, based on the training of growing pyramidal networks – a special data structure in the computer memory, which accelerates the search for classifying regularities [13]. In 1974, 239 examples of the ABO_3 composition compounds formation and 39 systems without compounds formation in this composition were selected for machine learning. Based on the physicochemical concepts, the following attributes sets were proposed to describe the A-B-O systems:

I - electrons distribution over the energy shells of isolated atoms and formal valences of elements A and B in ABO_3 compounds;
II - first four ionization potentials, incomplete electron shell types (s, p, d, f), ionic radii according to Bokii and Belov, standard isobaric heat capacities and formal valences of elements A and B in ABO_3 compounds;
III - standard heats of formation and standard isobaric heat capacities of the corresponding simple oxides, ionic radii according to Bokii and Belov of cations in oxides and formal valences of elements A and B in corresponding oxides.

To improve the calculations accuracy, predictions using different attributes sets were compared, and a decision on the new object status was made if the results did not contradict each other.

Over the past years, many of these composition compounds have been synthesized and it became possible to compare our predictions with new experimental data (Table 1).

Table 1. Part of the prediction table for $A^{2+}B^{4+}O_3$ composition compounds [2]

B^{4+} \ A^{2+}	Be	Mg	Ca	Mn	Fe	Co	Ni	Cu	Zn	Sr	Pd	Cd	Sn	Ba	Hg	Pb	Ra
C	↔	⊕	⊕	⊕	⊕	⊕	⊕	⊕	⊕	⊕	+	⊕	©	⊕	©	⊕	⊕
Si	©	⊕	⊕	⊕	⊕	©	©		⊕	⊕	+	⊕	©	⊕	+	⊕	+
S			©	©	©	©	+		©	©	+	©	©	©	©	©	+
Ti	↔	⊕	⊕	⊕	⊕	⊕	⊕	©	⊕	⊕	©	⊕	©	⊕	⊕	⊕	+
V			⊕	©	+	⊕	⊕	⊕	+	©	+	⊕	+	©	©	©	+
Mn	+	©	⊕	+	©	⊕	©	©	©	⊕	+	©	©	⊕	©	©	+
Ge	+	⊕	⊕	⊕	⊕	©	+	⊕	©	⊕	+	⊕	+	⊕	+	⊕	+
Se		⊕	©	©	©	©	©	©	©	©	©	©	©	©	©	©	+
Zr	↔	↔	⊕	+	↔	+	+	+	©	⊕	+	⊕	+	⊕	+	⊕	+
Mo		⊕	©	©	⊕	©	©	+	©	⊕	+	+	+	⊕	+	+	+
Tc			©	+	+	+	+	+	+	©	+	©	+	©	+	©	+
Ru			©	+	+	+	+	+	+	⊕	⊗	+	+	⊕	+	©	+
Sn			⊕	©	+	⊕	©	⊗	⊕	⊕	+	⊕	+	⊕	©	⊕	+
Te		⊕	⊕	⊕	+	©	⊕	⊕	⊕	⊕	+	©	©	©	©	©	+
Ce	↔	⊕	⊕					+	+	⊕	⊗	⊕		⊕	+	⊕	⊕
Pr								+	+	©	⊗	+		⊕	+	+	+
Tb	-	-	+					+	+	©	+	+		⊕			
Hf		↔	⊕	+	+	+	↔	+	+	⊕	+	©	©	⊕		⊕	
Ta			+	+	+	+	+	©	+	+	+	+	©	+			
W		+	©	+	+	+	+	+	+	©	+	+	+				
Re	-	-	+	+	+	+	+	+	+	+	+	+	+	©			
Os	-	-	⊕	+	+	+	+	+	+	⊕	+	©	+	⊕			
Ir			⊕	+	+	+	+	+	+	⊕	+	+	+	©			
Pt	-	-	©	+	+	+	+	+	+	+	+	+	+				
Pb			⊕	©	©	+	©	©	⊕	⊕	+	⊕	©	⊕	⊕		
Po			+	+	+	+	+	+	+	©	+	+	+	©			
Th	↔	⊕	⊕	+	+	+	+	+	+	⊕	+	⊕	+	⊕		⊕	
U	↔	↔	©	+	+	©	+	+	+	©	+	⊕	+	⊕			

The following designations are accepted: + - prediction of ABO_3 composition compounds formation under normal conditions; − - prediction of the absence of ABO_3 composition compounds in the A-B-O system under normal conditions; ⊕ - the ABO_3 composition compound exists and this fact is used for machine learning; ↔ - the ABO_3 composition compound is not formed in the A-B-O system and this fact is used for machine learning; © - the prediction of the ABO_3 composition compound formation is confirmed by experiment; O - the prediction of the ABO_3 composition compound absence in the A-B-O system is confirmed by experiment; ⊗ - the prediction of the ABO_3 composition compound is not confirmed by experiment; ∅ - the prediction of the ABO_3 composition compound absence in the A-B-O system is not confirmed by experiment; empty cells correspond to prediction mismatches for different attributes sets. From the 98 verified predictions for these compositions, only 4 were not confirmed experimentally, thus the prediction error is about 4%.

3.2 Predicting the Compounds Crystal Structure Type

Much more often, using ML methods, the compounds crystal structure type prediction problems (as a rule, under normal conditions) are solved, as well as problems related to the crystal structure, for example, the space group prediction [14], the octahedral tilting in oxide perovskites [15], etc.

An example of such solved problems is the ABO_3 composition perovskites prediction, which we carried out in the 70s [2]. In [2], not only compounds with a $CaTiO_3$ type cubic structure were assigned to the perovskites class, but also perovskites with a distorted ideal lattice, for example, orthorhombic perovskite-like compounds with the $GdFeO_3$ type structure. To describe the A-B-O systems in [2], attribute sets II and III were used (see Sect. 2.1). The training sample included 133 perovskites examples and 56 examples of systems in which these phases were not formed. The program [13] was used for machine learning. Table 2 shows the perovskites prediction comparison results with new experimental data. From the 68 experimentally verified predictions, 12 turned out to be incorrect, thus the error was less than 18%.

Table 2. Part of the prediction table for $A^{2+}B^{4+}O_3$ composition compounds with perovskite crystal structure [2]

A^{2+} / B^{4+}	Mg	Ca	Ti	V	Mn	Fe	Co	Ni	Cu	Zn	Ge	Sr	Pd	Cd	Sn	Ba	Hg	Pb
C	↔	↔			↔	↔	↔	↔	↔	↔		↔	-	↔	-	↔		↔
Si	↔	↔			↔	↔	O	O	↔		-	↔	-	O	-	O		↔
S		O			O	O	-	-	O		+	-	O		-	O		
Ti	↔	⊕	+	+	↔	↔		↔			+	⊕		⊕	©	⊕	⊕	⊕
V		⊕				-	↔	↔	↔		-	-	©	-				©
Mn	O	⊕	-	-	-	-	↔	↔	-	O	+	⊕	-	-	⊗	⊕	-	
Ge	↔	↔	-	-	↔	O	O	-	O	O	+	↔	-		+	↔	O	↔
Se	⊕	-			©	+	©	©	©	©	+		O	O		O	O	⊗
Zr		⊕		+		-	+		-			⊕		⊕		⊕		⊕
Nb					↔	↔	-	O	-			©	-					+
Mo	↔		+		↔	O	+	-	O			⊕	-	-	+			+
Tc					-	-	-	-	-			©	-	O				⊗
Ru					-	-	-	-	-			⊕	∅	-				⊗
Sn		↔	⊕	⊕		↔	∅		∅	+		⊕				⊕	⊗	↔
Te	⊕	O			©	⊗	©	©	⊕	O	O	+	⊗	-	O	-	O	
Ce	⊕	⊕					-	-				⊕	-	⊕		⊕		⊕
Pr							-					©	-	-		⊕	-	+
Tb							-	-				©	-			⊕		
Hf		⊕		+		+		+	-	+		⊕				⊕		⊕
Ta		-	-	-	-	-	-	-	O	-	+		-	-	∅	-		
Os		⊕	+	+	+			+		-		⊕			+	⊕		
Ir		⊕	+	+	+				+			⊕			+	©		
Pb		↔			-	O	O	O	-	O	+	⊗	-	↔	-	⊕	↔	
Th	⊕	⊕			-	-	-	-	-			⊕	-	⊕		⊕		⊕
U					-	-		-			+	©	-	-		⊕		
Np					-	-	-	-	-		+	©	-	-		©		
Pu				⊕	-	-	-	-	-		+	©	-	-		⊕		
Am	+	+	+	+	+	+	+	+	-	-	+	©	+	+	+	©	+	

In Table 2, the following designations are accepted: + - prediction of the perovskite structure; − - prediction of the perovskite structure absence; ⊕ - the compound has a perovskite structure and this fact is used for machine learning; ↔ - the compound has a structure that is different from the perovskite one and this fact is used for machine learning; © - the prediction of the compound formation a with a perovskite structure is confirmed by an experiment; O - prediction of a structure other than perovskite, confirmed by an experiment; $\otimes$ - the prediction of the perovskite structure is not confirmed by an experiment; $\oslash$ - prediction of a structure other than perovskite, not confirmed by an experiment; empty cells correspond to prediction mismatches for different attributes sets.

Naturally, these very first experiments in the machine learning methods application were still far from the modern level. Taking into account many years of experience in the inorganic compounds computer design, more advanced methods and programs for machine learning have been created, as well as large DB PISM and additional methods and programs that increase the prediction accuracy. Now the system for predicting the categorical compounds properties [4] includes the most popular pattern recognition programs based on: the algorithm for calculating estimates, binary decision trees method, Fisher's linear discriminant, the search for classes logical regularities, the search for two-dimensional linear separators, the linear machine algorithm, different versions of neural networks learning, the k-nearest neighbors method, deadlock test algorithm, genetic algorithm, support vector machine, statistically weighted voting, concept formation using growing pyramidal networks, etc.

The most promising means of improving prediction accuracy include collective decision-making methods. The IAS developed by us contains several programs with such methods [4]: Bayesian method, clustering and selection, decision templates, dynamic Woods method, complex committee methods, logical correction, convex stabilizer, generalized polynomial and algebraic correctors.

To estimate the prediction accuracy in the IAS, two methods are used: cross-validation and examination recognition of objects that are not contained in the training sample. Such a peculiarity of inorganic chemistry problems as the asymmetry of the training sample class sizes is a problem in the ML algorithms accuracy estimation. The examination recognition generalized error does not reflect the prediction errors for small-sized classes, therefore, for analyzing different algorithms prediction quality, it is promising to use ROC-curves, which make it possible to compare the recognition accuracy of the target and alternative classes when varying the thresholds that determine belonging to different classes. It should be emphasized that the most accurate algorithms are selected and used to predict the not yet obtained compounds.

One of the most successful ways to improve the predicting accuracy in the case of samples containing K classes objects ($K > 2$) is a multistage learning. First, multiclass learning and prediction are performed. Then K dichotomies are carried-out: the target class versus all alternative ones, followed by prediction. The results of multiclass prediction and a series of dichotomies are compared, and a decision on the object status is made if the predictions do not contradict each other. To solve this problem, software tools have been developed for making a collective solution based on comparing the results of multiclass prediction and a series of dichotomies.

An important subsystem of the IAS developed by us is a programs set for selecting the most informative attributes. It includes three programs based on algorithms [16–18]. In addition to the initial component parameters set, it is possible to automatically generate simple algebraic functions, which include initial parameters combinations, with the subsequent selection of the most informative functions. The IAS also includes a visualization subsystem, which makes it possible to view the points location corresponding to different training sample objects, by means of projections in user-specified coordinates. However, it should be noted that the use of only informative attributes when describing objects always reduces the prediction accuracy.

For expert evaluation of data for machine learning, programs have been developed [19, 20] for determining outliers in the data, which are associated with errors in assigning a substance to a certain class or errors in the attribute value (values) used to describe the substance. This toolkit significantly reduces and simplifies the work for specialists who select objects for machine learning.

To fill in the gaps in the values of the components properties, interpolation is used taking into account the elements properties periodic dependencies on other elements parameters. In this case the ML methods application possibility was also tested for quantitative properties prediction.

Based on our results comparison with new experimental data, the ML methods application provided an average accuracy of above 80% for the inorganic substances' qualitative properties prediction.

3.3 Quantitative Properties Prediction

Categorical properties prediction is just a small part of the practical problems in the inorganic compounds design. The overwhelming majority of problems are associated with the objects' quantitative characteristics prediction (for example, crystal lattice parameters, melting or boiling points, band gap, impact toughness, elasticity, electrical conductivity, etc.). To solve such problems, we have developed the ParIS test version (Parameters of Inorganic Substances) system, which includes specialized ML programs [5], taking into account the subject area peculiarities. In particular, the ParIS system includes well-known programs (methods based on the ridge regression, LASSO, LARS, elastic networks, regularized neural networks, etc.) from the scikit-learn package for Python programming language [9], which use regularization, various methods for the most important attributes selection, filtering outliers, etc. One of the promising approaches in the ML methods development for predicting the inorganic substances quantitative properties is the combined algorithms creation, for example, multilevel methods (for example, algorithm [21]), in which at each level an algorithms family is generated that predict the output value, while the lower level algorithms outputs (predictors) are fed to the input for a higher level algorithms. For example, the multilevel method application made it possible to almost double the values accuracy for the chalcospinels crystal lattice parameters estimation [22].

The subsystem for the machine learning quality estimation in the ParIS system allows to estimate the Mean Absolute Error (MAE) and Mean Squared Error (MSE) (with cross-validation in LOOCV mode - Leave-One-Out Cross-Validation), determination coefficient R^2 and etc., and also to construct the target parameters calculated values deviation diagram from the experimental ones for substances, information about which was used in machine learning. The deviation diagram also allows to identify outliers in the experimental data used for machine learning. To do this, the diagram provides output for experimental and predicted values for points marked by the user. For example, when analyzing such a diagram (Fig. 1), obtained during the lattice parameters examination recognition of double cubic perovskites with $A^{II}_2 B^{III} B'^{V} O_6$ composition, two outlier points corresponding to the Ba_2TiPuO_6 and Ba_2FeUO_6 compounds were analyzed. For the Ba_2TiPuO_6 compound, in the appendix to review [23], two values for the crystal lattice parameters are given: a = 8.06 and 8.87 Å. The predicted value is 8.26 Å. The publication, from which the value was taken [24], analysis, showed that an error was made in the review [23]. The correct value for the parameter is a = 8.06 Å [24]. The handbook [25] for cubic perovskite Ba_2FeUO_6 indicates the lattice parameter value a = 7.5 Å. The predicted value is 8.28 Å. In [26], the value a = 8.312 Å was measured for this compound. The training sample correction carried out by an expert in this way made it possible to increase the subsequent prediction accuracy.

The estimation of the crystal lattice parameters values in combination with the space group prediction of not yet obtained compounds is of great importance for subsequent quantum mechanical calculations, since it becomes possible to determine the atoms arrangement in the crystal lattice. Table 3 shows a part of the space group predictions and lattice parameters of double cubic perovskites with $A^{II}_2B^{III}B'^{V}O_6$ composition. The lattice parameters estimation accuracy using the Automatic Relevance Determination (ARD) Regression algorithm MAE = ±0.07 Å.

Machine learning methods have been used to predict inorganic substances various properties, for example, the transition temperature to the superconducting state of HTSC [27], the inorganic substances heat capacity in the solid state [28], the cohesive energy, the melting point and lattice thermal conductivity of double and ternary compounds [29], band gap [30], etc.

Table 3. The cubic crystal lattice parameter prediction for new compounds with $A_2^{II}B^{III}C^{V}O_6$ composition (sp. gr. Fm(-)3m)

Compound	a, Å	Compound	a, Å	Compound	a, Å
Ca_2AlUO_6	7.90	Sr_2PmUO_6	8.53	Ba_2PmSbO_6	8.43
Ca_2ScUO_6	8.12	Sr_2TbUO_6	8.52	Ba_2PmWO_6	8.53
Ca_2GaUO_6	8.08	Sr_2TmUO_6	8.45	Ba_2PmMoO_6	8.44
Ca_2YUO_6	8.36	Sr_2BiUO_6	8.68	Ba_2PmUO_6	8.70
Ca_2PrUO_6	8.47	Sr_2AmUO_6	8.60	Ba_2GdRuO_6	8.35
Ca_2PmUO_6	8.40	Ba_2AlMoO_6	7.94	Ba_2TbRuO_6	8.34
Ca_2GdUO_6	8.39	Ba_2AlWO_6	8.03	Ba_2TbWO_6	8.51
Ca_2TbUO_6	8.38	Ba_2AlReO_6	7.87	Ba_2HoWO_6	8.49
Ca_2DyUO_6	8.36	Ba_2AlOsO_6	7.91	Ba_2TmVO_6	8.20
Ca_2HoUO_6	8.35	Ba_2AlUO_6	8.21	Ba_2TmWO_6	8.44
Ca_2ErUO_6	8.34	Ba_2ScVO_6	8.00	Ba_2TmUO_6	8.62
Ca_2TmUO_6	8.31	Ba_2ScWO_6	8.25	Ba_2YbVO_6	8.16
Ca_2YbUO_6	8.27	Ba_2VRuO_6	7.94	Ba_2YbWO_6	8.40
Ca_2LuUO_6	8.27	Ba_2VWO_6	8.11	Ba_2BiUO_6	8.85
Ca_2AmUO_6	8.47	Ba_2VUO_6	8.28	Ba_2AmNbO_6	8.55

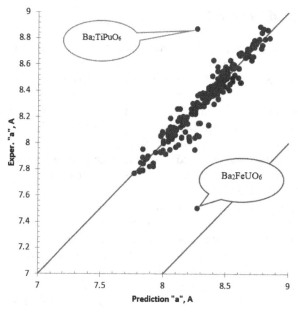

Fig. 1. Predicted lattice parameter values deviations diagram from experimental ones for cubic perovskites with $A^{II}_2B^{III}B^{\cdot V}O_6$ composition (ARD Regression).

3.4 Machine Learning Methods Application in Inorganic Materials Industry

Prediction of new inorganic materials with specified properties and technology optimization for their production and processing are the tasks most often solved by ML methods. The publications number in this area grows exponentially every year, covering a wide variety of areas: metallurgy (for example, works devoted to optimization of strength and ductility of cast magnesium alloys [31], prediction of thermoviscoplasticity during steels solidification [32], etc.), electronics (the controlling recipe optimization in quasi-single crystalline silicon growth [33], modeling the voltametric behavior of supercapacitors [34], etc.), the building materials industry (for example, studies related to the compressive strength prediction of lightweight foamed concrete [35]), nuclear power (for example, predicting the thermal conductivity of irradiated uranium-molybdenum nuclear fuels [36]), etc. One of the problems in solving these tasks is the lack of high-quality and representative data for machine learning. There is often a lot of data taken from automated industrial plants, but the quality is poor. In these cases, when preparing a training sample for computer analysis, a serious expert assessment is required for the available information. Nevertheless, the machine learning application to industry is very promising and will result in the inorganic materials production automation and robotization.

4 Problems and Prospects

Machine learning is a powerful tool for solving a wide variety of problems in inorganic chemistry and materials science. Our experience and the experience of other researchers allowed us to outline the prospects for the approach development as applied to this subject area, aimed at solving the problems that have arisen.

One of the most important problems is the quality and representativeness of information used for machine learning. The only way to solve this problem is to create large databases on inorganic substances and materials properties with high-quality information and conceptual structure, designed to search for information required for machine learning quickly. An example of such a DB PISM is the integrated information system developed at IMET RAS [3].

The laborious validation task of the DB PISM information, which requires highly qualified experts in the subject area, can be partially automated through various programs for determining outliers in data application: from a set of programs using statistics criteria (for example, [19, 20]), to deviations diagrams of the predicted parameter values from the experimental ones [5, 22].

The informative attributes selection for inclusion in the searched regularities heavily affects the prediction accuracy. In most cases, this process requires large computational resources, especially in those tasks in which numerous functions from the original attributes are added. The new programs development for solving such problems is one of the most promising areas of the ML. One of the ways to reduce the computational resource consumption in the categorical properties prediction problems is associated with the special data structures creation at the initial stage of entering the training sample, for example, as it was done in algorithms [13].

The machine learning quality assessment problem is still partially solved especially in tasks with different class sizes when predicting categorical properties and in problems with small samples for machine learning when assessing quantitative parameters. The ROC-curves use in the first case and examination recognition in the LOOCV mode in the second case, allows us to approximately determine the subsequent prediction error.

The new machine learning methods development, taking into account the subject area specifics, is the most important task when using these methods in inorganic chemistry and materials science. The most promising areas are the collective decision-making methods application (ML algorithms ensembles, multistage learning, etc.), as well as combined methods (neural networks deep learning [37], multilevel methods [21], etc.).

Acknowledgements. The authors are grateful to V.V. Ryazanov, O. V. Sen'ko, A.A. Dokukin, V.S. Pereverzev-Orlov, M.A. Vitushko, and E.A. Vaschenko for their help in developing algorithms and programs. This work was supported in part by the Russian Foundation for Basic Research, project nos. 20-01-00609 and 18-07-00080. The study was carried out as part of the state assignment (project no. № 075-00328-21-00).

References

1. Savitskii, E.M., Devingtal', Yu.V., Gribulya, V.B.: Prediction of metallic compounds with composition A_3B using computer. Dokl. Akad. Nauk SSSR **183**, 1110–1112 (1968). (in Russ.)
2. Kiselyova, N.N., Pokrovskii, B.I., Komissarova, L.N., Vaschenko, N.D.: Simulation of the complicated oxides formation from initial components based on the cybernetic method of concept formation. Russ. J. Inorg. Chem. **22**, 883–886 (1977). (in Russ.)
3. Kiselyova, N.N., Dudarev, V.A., Stolyarenko, A.V.: Integrated system of databases on the properties of inorganic substances and materials. High Temp. **54**, 215–222 (2016). https://doi.org/10.1134/S0018151X16020085
4. Kiselyova, N.N., Stolyarenko, A.V., Ryazanov, V.V., Sen'ko, O.V., Dokukin, A.A., Podbel'skii, V.V.: A system for computer-assisted design of inorganic compounds based on computer training. Pattern Recognit. Image Anal. **21**, 88–94 (2011). https://doi.org/10.1134/S10 54661811010081
5. Dudarev, V.A., et al.: An information system for inorganic substances physical properties prediction based on machine learning methods. In: CEUR Workshop Proceedings (CEUR-WS.org), vol. 2790. Supplementary Proceedings of the XXII International Conference on Data Analytics and Management in Data Intensive Domains (DAMDID/RCDL 2020), pp. 89–102 (2020). http://ceur-ws.org/Vol-2790/paper09.pdf
6. Site of Materials Genome Initiative. https://www.mgi.gov/. Accessed 30 Mar 2021
7. Site of Novel Materials Discovery Laboratory. http://nomad-lab.eu/. Accessed 30 Mar 2021
8. Site of Center for Materials Research by Information Integration. http://www.nims.go.jp/eng/research/MII-I/index.html. Accessed 30 Mar 2021
9. Site of scikit. http://scikit-learn.org/. Accessed 30 Mar 2021
10. Site of R. https://www.r-project.org/. Accessed 30 Mar 2021
11. Devingtal', Yu.V.: Coding of objects at application of separating hyper-plane for their classification. Izv. Akad. Nauk SSSR. Tekhn. Kibernetika. 139–147 (1971). (in Russ.)
12. Seko, A., Hayashi. H., Tanaka, I.: Compositional descriptor-based recommender system for the materials discovery. J. Chem. Phys. **148**, 241719/1-7 (2018). https://doi.org/10.1063/1.5016210
13. Gladun, V.P.: Heuristic Search in Complex Environments. Naukova Dumka, Kiev (1977).(in Russ.)
14. Liu, C.-H., Tao, Y., Hsu, D., Du, Q., Billinge, S.J.L.: Using a machine learning approach to determine the space group of a structure from the atomic pair distribution function. Acta Crystallogr. A **75**, 633–643 (2019). https://doi.org/10.1107/S2053273319005606
15. Xie, S.R., Kotlarz, P., Hennig, R.G., Nino, J.C.: Machine learning of octahedral tilting in oxide perovskites by symbolic classification with compressed sensing. Comp. Mater. Sci. **180**, 109690/1-9 (2020). https://doi.org/10.1016/j.commatsci.2020.109690
16. Senko, O.V.: An optimal ensemble of predictors in convex correcting procedures. Pattern Recognit Image Anal. **19**, 465–468 (2009). https://doi.org/10.1134/S1054661809030110
17. Yuan, G.-X., Ho, C.-H., Lin, C.-J.: An improved GLMNET for L1-regularized logistic regression. J. Mach. Learn. Res. **13**, 1999–2030 (2012)
18. Yang, Y., Zou, H.A.: Coordinate majorization descent algorithm for L1 penalized learning. J. Stat. Comput. Simul. **2014**(84), 1–12 (2014). https://doi.org/10.1080/00949655.2012.695374
19. Ozhereliev, I.S., Senko, O.V., Kiselyova, N.N.: Method for searching outlier objects using parameters of learning instability. Sist. Sredstva inform. – Syst. Means Inform. **29**, 122–134 (2019). https://doi.org/10.14357/08696527190211. (inRuss.)

20. Dineev, V.D., Dudarev, V.A.: Extendable system for multicriterial outlier detection. In: CEUR Workshop Proceedings (CEUR-WS.org), vol. 2790. Supplementary Proceedings of the XXII International Conference on Data Analytics and Management in Data Intensive Domains (DAMDID/RCDL 2020), pp. 103–113 (2020). http://ceur-ws.org/Vol-2790/paper10.pdf. (in Russ.)

21. Senko, O.V., Dokukin, A.A., Kiselyova, N.N., Khomutov, N.: Two-stage method for constructing linear regressions using optimal convex combinations. Dokl. Math. **97**, 113–114 (2018). https://doi.org/10.1134/S1064562418020035

22. Kiselyova, N.N., Dudarev, V.A., Ryazanov, V.V., Sen'ko, O.V., Dokukin, A.A.: Predictions of chalcospinels with composition $ABCX_4$ (X – S or Se). Inorg. Mater.: Appl. Res. **12**, 328–336 (2021). https://doi.org/10.1134/S2075113321020246

23. Vasala, S., Karppinen, M.: $A_2B'B''O_6$ perovskites: a review. Progr. Solid State Chem. **43**, 1–36 (2015). https://doi.org/10.1016/j.progsolidstchem.2014.08.001

24. Awasthi, S.K., Chackraburtty, D.M., Tondon, V.K.: Studies on $A_2BB'O_6$ type compounds of actinides: Plutonium compounds. J. Inorg. Nucl. Chem. **30**, 819–821 (1968). https://doi.org/10.1016/0022-1902(68)80442-7

25. Landolt-Bornstein. Zahlenwerte und Funktionen aus Naturwissenschaften und Technik. Neue Serie. Gr.III: Kristal- und Festkorperphysik. B.7. Kristallstrukturdaten anorganischer Verbindungen. T.e: Schlusselemente: d9-, d10-, d1...d3-, f-Elemente. Springer, Berlin, Heidelberg, New York (1976)

26. Sleight, A.W., Ward, R.: Compounds of hexavalent and pentavalent uranium with the ordered perovskite structure. Inorg. Chem. **1**, 790–793 (1962). https://doi.org/10.1021/ic50004a015

27. Torshin, I.Yu, Rudakov, K.V.: Topological data analysis in materials science: the case of high-temperature cuprate superconductors. Pattern Recognit. Image Anal. **30**, 264–276 (2020). https://doi.org/10.1134/S1054661820020157

28. Kauwe, S.K., Graser, J., Vazquez, A., Sparks, T.D.: Machine learning prediction of heat capacity for solid inorganics. Integr. Mater. Manuf. Innov. **7**(2), 43–51 (2018). https://doi.org/10.1007/s40192-018-0108-9

29. Seko, A., Hayashi, H., Nakayama, K., Takahashi, A., Tanaka, I.: Representation of compounds for machine-learning prediction of physical properties. Phys. Rev. **B99**, 144110/1-11 (2017). https://doi.org/10.1103/PhysRevB.95.144110

30. Lee, J., Seko, A., Shitara, K., Tanaka, I.: Prediction model of band gap for inorganic compounds by combination of density functional theory calculations and machine learning techniques. Phys. Rev. **B93**, 115104/1-12 (2016). https://doi.org/10.1103/PhysRevB.93.115104

31. Chen, Y., et al.: Machine learning assisted multi-objective optimization for materials processing parameters: a case study in Mg alloy. J. Alloys Compounds. **844**, 156159/1-7 (2020). https://doi.org/10.1016/j.jallcom.2020.156159

32. Abueidda, D.W., Koric, S., Sobh, N.A., Sehitoglu, H.: Deep learning for plasticity and thermo-viscoplasticity. Int. J. Plasticity. **136**, 102852/1-30 (2021). https://doi.org/10.1016/j.ijplas.2020.102852

33. Dang, Y., Liu, L., Li, Z.: Optimization of the controlling recipe in quasi-single crystalline silicon growth using artificial neural network and genetic algorithm. J. Crystal Growth. **522**, 195–203 (2019). https://doi.org/10.1016/j.jcrysgro.2019.06.033

34. Parwaiz, S., Malik, O.A., Pradhan, D., Khan, M.M.: Machine learning-based cyclic voltammetry behavior model for supercapacitance of co-doped Ceria/rGO nanocomposite. J. Chem. Inf. Model. **58**, 2517–2527 (2018). https://doi.org/10.1021/acs.jcim.8b00612

35. Yaseen, Z.M., et al.: Predicting compressive strength of lightweight foamed concrete using extreme learning machine model. Adv. Eng. Software. **115**, 112–125 (2018). https://doi.org/10.1016/j.advengsoft.2017.09.004

36. Kautz, E.J., Hagen, A.R., Johns, J.M., Burkes, D.E.: A machine learning approach to thermal conductivity modeling: a case study on irradiated uranium-molybdenum nuclear fuels. Comp. Mater. Sci. **161**, 107–118 (2019). https://doi.org/10.1016/j.commatsci.2019.01.044

37. Schmidhuber, J.: Deep learning in neural networks: an overview. Neural Netw. **61**, 85–117 (2015). https://doi.org/10.1016/j.neunet.2014.09.003

Interoperability and Architecture Requirements Analysis and Metadata Standardization for a Research Data Infrastructure in Catalysis

Martin Horsch[1,2]([✉])[iD], Taras Petrenko[1][iD], Volodymyr Kushnarenko[1][iD], Bjoern Schembera[1,3][iD], Bianca Wentzel[4][iD], Alexander Behr[5][iD], Norbert Kockmann[5][iD], Sonja Schimmler[4][iD], and Thomas Bönisch[1][iD]

[1] High Performance Computing Center Stuttgart (HLRS), Nobelstr. 19, 70569 Stuttgart, Germany
{taras.petrenko,volodymyr.kushnarenko,thomas.boenisch}@hlrs.de
[2] School of Psychology and Computer Science, University of Central Lancashire, 45 Fylde Road, Preston, Lancashire PR1 1JN, UK
mthorsch@uclan.ac.uk
[3] Institute of Applied Analysis and Numerical Simulation, University of Stuttgart, Allmandring 5b, 70569 Stuttgart, Germany
bjoern.schembera@ians.uni-stuttgart.de
[4] Fraunhofer Institute for Open Communication Systems (FOKUS), Kaiserin-Augusta-Allee 31, 10589 Berlin, Germany
{bianca.wentzel,sonja.schimmler}@fokus.fraunhofer.de
[5] Department of Biochemical and Chemical Engineering, Laboratory of Equipment Design, TU Dortmund University, Emil-Figge-Str. 68, 44227 Dortmund, Germany
{alexander.behr,norbert.kockmann}@tu-dortmund.de

Abstract. The National Research Data Infrastructure for Catalysis-Related Sciences (NFDI4Cat) is one of the disciplinary consortia formed within the German national research data infrastructure (NFDI), an effort undertaken by the German federal and state governments to advance the digitalization of all scientific research data within the German academic system in accordance with the FAIR principles. This work reports on initial outcomes from the NFDI4Cat project. The data value chain in catalysis research is analysed, and architecture and interoperability requirements are identified by conducting user interviews, collecting competency questions, and exploring the landscape of semantic artefacts. Methods from agile software development are employed to collect, organize, and present the collected requirements; workflows are annotated on the basis of metadata standards for research data provenance, by which requirements for domain ontologies in catalysis are deduced.

Keywords: Research data infrastructure · Applied ontology · Catalysis

© Springer Nature Switzerland AG 2022
A. Pozanenko et al. (Eds.): DAMDID/RCDL 2021, CCIS 1620, pp. 166–177, 2022.
https://doi.org/10.1007/978-3-031-12285-9_10

1   Introduction

The German national research data infrastructure (NFDI) is a long-term initiative supported by the German federal and state governments that targets the digitalization of all scientific research data. It constitutes itself as a public-benefit association (NFDI e.V. [26,40]) and is advanced through consortia funded from a series of calls that are managed by the German Research Foundation (DFG). In this way, up to 30 disciplinary consortia will be established [55]. Within the two initial stages of funding, 19 projects have been approved; consortia covering domains of knowledge relevant to the topic of the present DAMDID workshop, *data and computation for materials science and innovation* (DACOMSIN), include FAIRmat (condensed-matter and solid-state physics), MaRDI (mathematics [23]), NFDI4Cat (catalysis-related sciences [16,53,60]), NFDI4Chem (chemistry [27–29]), NFDI4DataScience, which addresses the eponymous field, NFDI4Ing (engineering sciences [10]), and NFDI-MatWerk (materials science and engineering [44]).

At the intersection of process systems engineering, technical and theoretical chemistry, materials science, and biotechnology, the *National Research Data Infrastructure for Catalysis-Related Sciences* (NFDI4Cat) develops a federated repository architecture for data on catalysts and their technical applications [16,53,60]. The first funding period of the NFDI4Cat project started in October 2020 and extends over five years. This work reports on initial outcomes from NFDI4Cat, with a focus on requirements for metadata standardization, data management practice, and the repository architecture. It is structured as follows: Sect. 2 introduces the methodology employed for the present requirements analysis. Section 3 comments on research data provenance documentation and aspects of the research workflows that were discussed with project stakeholders in detail, grounding the present requirements analysis. In Sect. 4, the landscape of pre-existing ontologies in the relevant domains is surveyed, following the data value chain in catalysis and process industry, and challenges for future work within the NFDI4Cat project are summarized. Section 5 summarizes the method that NFDI4Cat will employ for ontology design, extension, and population. Requirements for the repository architecture are discussed in Sect. 6. The conclusion in Sect. 7 outlines major next steps and future perspectives.

2   Requirements Analysis Methodology

Most NFDI consortia have regarded it as advisable to conduct a requirements analysis before undertaking major development efforts; *e.g.*, NFDI4Chem [29], NFDI4Ing [36], and NFDI-MatWerk [44] have followed questionnaire-based approaches to collect information on the state of the art and the priorities within their respective disciplinary communities. Typical questions that were addressed by written communication include "which data are generated where?," "what kinds of storage facilities are available?," and "are data annotated and is provenance documented for future re-use?," *cf.* Herres-Pawlis *et al.* [29]. In NFDI4Cat, to elicit requirements for the repository architecture and data management practice, 21 interviews with researchers from consortial partners and associated institutions were conducted, and three methods with a focus on direct person-to-person communication were applied to evaluate the user interviews [53].

First, following agile software development practice as outlined by Cohn [13], requirements identified at user-interview stage (and, subsequently, by exchanging minutes and detailed interview documentation material) are documented in terms of *epics* and *user stories*. There, a user story "describes functionality that will be valuable to" an end user [13]. Good user stories are *independent, negotiable, valuable, estimable, small,* and *testable* (INVEST) [11]. These requirements are connected hierarchically such that multiple user stories elucidate concrete aspects or parts of an epic [13]; accordingly, user stories that share a common motivation are grouped into an epic, *i.e.*, a coarse-grained objective. When in doubt, the granularity level distinction between epics and user stories can be made on the basis of metrics such as the expected implementation duration [41]. Each user story and epic is associated with a *persona*, *i.e.*, "an imaginary representation of a user role" [13]. These fictitious representative user identities are employed at documentation stage instead of the real names of the interview partners so that end users with similarly aligned concerns can be grouped together; moreover, the use of pseudonyms encourages users to voice their concerns more freely. The user stories are then stated as succinctly as possible, comprising three parts: A role description, a task, and a higher-order objective; *e.g.*, a user story from the point of view of a Scientific Data Officer [51] is given as follows:

As: Scientific Data Officer,   **I intend to:** Express a quantitative (vote) and/or qualitative (comment) evaluation of published results, **subordinate to my aim:** To enable peer review of data.

Second, requirements for data documentation are obtained by collecting *competency questions*, a widespread method in ontology design [7,20] introduced by Grüninger and Fox [25]. Competency questions are requests for information, formulated in any applicable register of human language (from colloquial to disciplinary scientific) that *a)* need to be expressible as SPARQL queries [59] on the basis of the employed ontologies [2, therein, Tab. 12] and that *b)* the research data infrastructure needs to be capable of addressing competently by facilitating appropriate modes of data ingest, retrieval, and extraction.

Third, during the interviews, representative *research workflows* were discussed in detail; some aspects of these workflows are summarized in Sect. 3.

3   Research Workflows and Data Provenance

The collected research workflows are one of the points of departure for NFDI4Cat ontology development, since the *provenance of research data* is most completely documented by describing the research workflow by which they are generated. Therefore, any domain-specific concepts and relations that appear in these workflow descriptions, but are missing from pre-existing semantic artefacts, constitute requirements for domain-ontology development (see Sect. 4). Simulation driven research relies on data and models at multiple granularity levels (highlighted in green circles), extending from the *electronic* level over *atomistic and mesoscopic* modelling up to *continuum* granularity-level process simulation. Research data provenance at these granularity levels can be documented with OSMO [34,35],[1] the ontology version of the CEN workshop agreement MODA [17], for modelling and simulation workflows; the provenance of experimental characterization data can be documented through CHADA [48]. For general business process models, the BPMN ontology [49], a PM2ONTO based ontology [47], and the BBO [2] are available as ontologizations of the BPMN standard [1,38]. Addressing both experimental and simulation research workflows as well as decision making processes more generally as *cognitive processes*, PIMS-II [32,39] can be employed as a mid-level ontology[2] that is aligned with the Elementary Multiperspective Material Ontology (EMMO) [12,21,30,33].

4   Semantic Interoperability

For research and development with industrial applications, which includes research on catalysis, enhancing the data value chain, *i.e.*, the data-driven aspects of the industrial value-added chain, becomes a priority for disciplinary research data management in general, including all work carried out in the academic sector. Accordingly, the data value chain of catalysis as depicted in Fig. 1 is used as a guideline to evaluate and match pre-existing ontologies in the field of catalysis.

Three domains of knowledge are targeted for formalization by domain ontologies to be developed within NFDI4Cat: First, applications where catalysis data are relevant; this consists of the subcategories synthesis, operando, performance, and characterization data. Here, the focus lies on the characterization of catalysts at the molecular level with an orientation toward materials science and engineering. The second domain is represented by applications where typical parameters are identified for heat and mass transfer, kinetic data of catalysed reactions, and the associated catalyst properties. The third domain is concerned with concepts regarding process engineering, chemical equipment, and plant design as well as catalytic process applications. Irrespective of the domain, it is essential to be able to document data provenance (see the discussion and the ontologies listed in Sect. 3). Over 40 semantic artefacts, mostly ontologies, were evaluated regarding their relevance to these domains. For active sites, covering catalyst synthesis and operando data, *e.g.*, OntoCompChem [18], OntoKin [18,19], and RXNO [43] were identified. Catalyst characterization and performance data can be formalized, *e.g.*, by OntoCompChem [18], AniML [24, Section 62], and EnzymeML [46]. The system of EMMO domain ontologies, *cf.* Clark *et al.* [12], covers aspects of synthesis, characterization, and use cases relevant to catalysis; in particular,

---

[1] VIMMP ontologies [31]: https://gitlab.com/vimmp-semantics/vimmp-ontologies/.
[2] Ontology accessible at http://www.molmod.info/semantics/pims-ii.ttl.

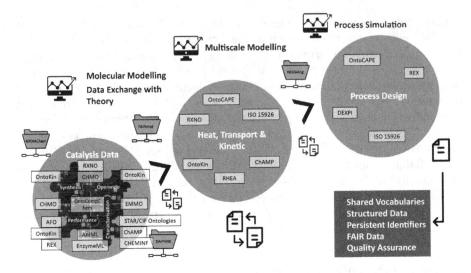

Fig. 1. Landscape of semantic artefacts relevant to the data value chain of catalysis.

the BattInfo ontology and BVCO are available for metadata related to electro-catalysis and battery research [12]. Regarding heat and mass transfer as well as reaction kinetics, candidate ontologies include OntoCAPE [58], OntoKin [18], and the ISO 15926 ontology [8,9]. DEXPI [56], the ISO 15926 ontology [8,9], and OntoCAPE [58] were identified as relevant semantic artefacts for process and product design. All relevant semantic artefacts are collected on the NFDI4Cat website[3], which is continuously updated. Further, NFDI4Ing has released its overarching ontology Metadata4Ing[4] [42], on the basis of previous efforts including EngMeta[5] [52], which is designed for engineering sciences, but has some topical overlap with the ongoing metadata standardization work from NFDI4Cat.

5   Research Workflow to Ontology Enhancements

The landscape of semantic artefacts for catalysis research is quite heterogeneous (*cf.* Fig. 1). While there are many artefacts present, an ontology that addresses all the needs of catalysis research does not exist so far. Thus, a workflow is proposed to extend existing ontologies by domain knowledge relevant to catalysis research. In a first step, terms to describe a concept are extracted from text sources such as, *e.g.*, scientific papers. Domain experts can review these concept terms and add missing ones to create a curated list of concepts important to the description of the respective domain knowledge. Ontology analysis is then

---

[3] The collection of relevant semantic assets can be found on https://nfdi4cat.org/en/services/ontology-collection/.

[4] Documentation and ontology accessible at https://nfdi4ing.pages.rwth-aachen.de/metadata4ing/metadata4ing/.

[5] XML schema definition accessible at https://dx.doi.org/10.18419/darus-500.

required to identify the correct ontology to be extended by the concepts gathered. Here, a validation of the contained concepts also takes place, identifying any remaining gaps to be closed. With ontologies presented in Fig. 1, an ontology database can be set up, and entries from these ontologies can be explored. Examples for possible environments to set up such an open-source ontology database with ontologies relevant to catalysis are the Ontology Lookup Service (OLS) [37] and Bioportal [57]. Using such an environment enables fast search and exploration of ontologies by users. Also, queries can be automated by APIs. Once the ontology database is set up and the pre-existing ontologies have been extended as required, research data and metadata can be processed and mapped to the respective ontologies. This allows for the population of ontologies, resulting in knowledge graphs that are, *e.g.*, suitable for ingest into a FAIR knowledge base. This overall workflow to extend ontologies is visualized in Fig. 2.

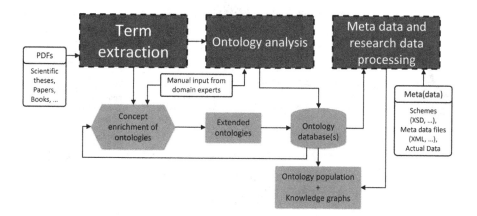

Fig. 2. Workflow for ontology extension and population.

6   Repository Architecture

In collaboration with the interview partners, based on the user interviews with which fictitious personas are associated as pseudonyms [13], specific epics and user stories are formulated as requirements for the infrastructure. The epics and user stories are grouped into the following categories: Metaportal (MP); repository (GR); repository harvester (RH); storage harvester (SH); data security (DS); metadata standards (MD); non-functional requirements (NF).

The totality of these user stories serves as a reference point for developing and implementing an infrastructure as shown in Fig. 3, covering the whole research data lifecycle. At the different sites, local tools are available for data collection, data processing, and data analysis. These tools interact with the overarching data infrastructure, which is in turn divided into three main components (*cf.* Fig. 3).

At the bottom, there is a layer to enable distributed data sources, where needed (distributed storage layer). Matching institutional requirements for confidentiality, some or all of these data are preserved in the repository layer, either in a local repository with access restrictions or in an overarching repository (repository layer). All repositories are connected to the metaportal (presentation layer) serving as the main data access point and single point of entry for external users. Because of the wide range of disciplines in catalysis research, there are specific needs and requirements regarding the infrastructure, depending on the field and methodology. Furthermore, because of the large involvement of industrial partners in catalysis, there is a need for an extensive permission management strategy, particularly for the overarching infrastructure, but also within single institutions.

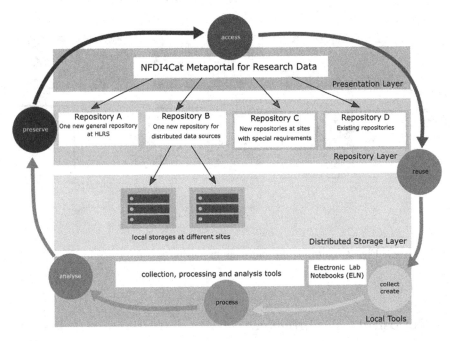

Fig. 3. Components of the federated NFDI4Cat repository architecture.

7   Conclusion

Design objectives and requirements identified by user-centered analysis serve as guidelines for developing the overarching and local repositories, domain ontologies, interfaces, and tools to be implemented as components of NFDI4Cat; as major next steps, available software packages will be evaluated for a potential reuse within NFDI4Cat, and pilot repositories will be built to establish the viability of the approach. Simultaneously, NFDI4Cat is actively working toward a training programme for disciplinary research data management to support the development of competencies and the community uptake of implemented solutions beyond the group of consortial members, so that it can accomplish its purpose of enabling all German academic institutions to digitalize their data on catalysis and process technology. Overall, the FAIR digitalization of all research data, as envisioned by the NFDI programme, will entail a paradigm-shifting innovation of research practices in ways that cannot yet be reliably predicted in detail; the present requirements analysis for NFDI4Cat is an attempt at such a prediction on the basis of momentary stakeholder expectations. As developers and users remain in contact, jointly reflecting on the platform and the NFDI at large, requirements and requests for additional functionality will further develop. Beside supporting the construction of a research data infrastructure for catalysis-related sciences, these efforts promise to advance process analytical technology in catalysis at large [6,15,50]; e.g., by facilitating the combination of data from different sources, covering widespread experimental techniques and supporting the integration of experiment, simulation, and data analytics into a coherent framework. This has the potential to finally bring into existence, at a large scale, the long envisioned digital interfaces between experiment and simulation as well as process and product design [4,5]. Accordingly, there will be synergies from reusing outcomes from this work for designing physical and digital twins in the context of cyber-physical production systems [3,14,15,22,45,54].

Acknowledgment. This work was funded by DFG through NFDI4Cat, DFG project no. 441926934, within the NFDI programme of the Joint Science Conference (GWK).

References

1. Allweyer, T.: BPMN 2.0: Introduction to the Standard for Business Process Modeling, 2nd edn. BoD, Norderstedt (2016). ISBN 978-3-8370-9331-5
2. Annane, A., Aussenac-Gilles, N., Kamel, M.: Une ontologie des processus métier (BBO) pour guider un agent virtuel. In: Hernandez, N. (ed.) Proceedings of IC 2019, pp. 183–198. AFIA (2019). https://hal.archives-ouvertes.fr/hal-02284535
3. Appl, C., Baganz, F., Hass, V.C.: Development of a digital twin for enzymatic hydrolysis processes. Processes **9**(10), 1734 (2021). https://doi.org/10.3390/pr9101734
4. Asprion, N., et al.: INES: an interface between experiments and simulation to support the development of robust process designs. Chem. Ing. Tech. **87**(12), 1810–1825 (2015). https://doi.org/10.1002/cite.201500020
5. Asprion, N., et al.: INES: interface between experiments and simulation. Comput. Aided Chem. Eng. **33**, 1159–1164 (2014). https://doi.org/10.1016/B978-0-444-63455-9.50028-3
6. Bai, J., Cao, L., Mosbach, S., Akroyd, J., Lapkin, A.A., Kraft, M.: From platform to knowledge graph: evolution of laboratory automation. JACS Au (2022). https://doi.org/10.1021/jacsau.1c00438
7. Barbosa Fernandes, P.C., Guizzardi, R.S.S., Guizzardi, G.: Using goal modeling to capture competency questions in ontology-based systems. J. Inf. Data Manag. **2**(3), 527–540 (2011)

8. Batres, R.: Ontologies in process systems engineering. Chem. Ing. Tech. **89**(11), 1421–1431 (2017). https://doi.org/10.1002/cite.201700037

9. Batres, R., et al.: An upper ontology based on ISO 15926. Comput. Chem. Eng. **31**(5–6), 519–534 (2007). https://doi.org/10.1016/j.compchemeng.2006.07.004

10. Bronger, T., Demandt, É., Heine, I., Kraft, A., Preuß, N., Schwarz, A.: Die Nationale Forschungsdateninfrastruktur für die Ingenieurwissenschaften (NFDI4Ing). Bausteine Forschungsdatenmanagement **2021**(2), 110–123 (2021). https://doi.org/10.17192/bfdm.2021.2.8329

11. Buglione, L., Abran, A.: Improving the user story agile technique using the INVEST criteria. In: Demirors, O., Can, A.B., Eşmelioğlu, S. (eds.) Proceedings of ISWM-MENSURA 2013, pp. 49–53. IEEE (2014). ISBN 978-0-7695-5078-7

12. Clark, S., et al.: Toward a unified description of battery data. Adv. Energy Mater. (2022). https://doi.org/10.1002/aenm.202102702

13. Cohn, M.: User Stories Applied for Agile Software Development. Pearson Education, Boston (2004). ISBN 978-0-321-20568-1

14. Deagen, M.E., Brinson, L.C., Vaia, R.A., Schadler, L.S.: The materials tetrahedron has a "digital twin". MRS Bull. (2022). https://doi.org/10.1557/s43577-021-00214-0

15. Eifert, T., Eisen, K., Maiwald, M., Herwig, C.: Current and future requirements to industrial analytical infrastructure—part 2: smart sensors. Anal. Bioanal. Chem. **412**(9), 2037–2045 (2020). https://doi.org/10.1007/s00216-020-02421-1

16. Espinoza, S., et al.: NFDI for catalysis-related sciences. Bausteine Forschungsdatenmanagement **2021**(2), 57–71 (2021). https://doi.org/10.17192/bfdm.2021.2.8333

17. European Committee for Standardization: Materials Modelling: Terminology, Classification and Metadata. CEN Workshop Agreement (expired) 17284:2018 (E), CEN (2018). https://www.cencenelec.eu/media/CEN-CENELEC/CWAs/RI/cwa17284_2018.pdf, validity expired (the validity of this agreement extended until 17th April 2021, three years after its publication)

18. Farazi, F., et al.: Linking reaction mechanisms and quantum chemistry: an ontological approach. Comput. Chem. Eng. **137**, 106813 (2020). https://doi.org/10.1016/j.compchemeng.2020.106813

19. Farazi, F., et al.: OntoKin: an ontology for chemical kinetic reaction mechanisms. J. Chem. Inf. Model. **60**(1), 108–120 (2020). https://doi.org/10.1021/acs.jcim.9b00960

20. Fernández-Izquierdo, A., García-Castro, R.: Requirements behaviour analysis for ontology testing. In: Faron Zucker, C., Ghidini, C., Napoli, A., Toussaint, Y. (eds.) EKAW 2018. LNCS (LNAI), vol. 11313, pp. 114–130. Springer, Cham (2018). https://doi.org/10.1007/978-3-030-03667-6_8. ISBN 978-3-030-03666-9

21. Francisco Morgado, J., et al.: Mechanical testing ontology for digital-twins: a roadmap based on EMMO. In: García Castro, R., Davies, J., Antoniou, G., Fortuna, C. (eds.) Proceedings of SeDiT 2020, p. 3. CEUR-WS (2020). http://ceur-ws.org/Vol-2615/

22. Glatt, M., Sinnwell, C., Yi, L., Donohoe, S., Ravani, B., Aurich, J.C.: Modeling and implementation of a digital twin of material flows based on physics simulation. J. Manuf. Syst. **58B**, 231–245 (2021). https://doi.org/10.1016/j.jmsy.2020.04.015

23. Görgen, C., Sinn, R.: Mathematik in der Nationalen Forschungsdateninfrastruktur. Mitteilungen der Deutschen Mathematiker-Vereinigung **29**(3), 122–123 (2021). https://doi.org/10.1515/dmvm-2021-0049

24. Gressling, T.: Data Science in Chemistry. De Gruyter, Berlin (2021). ISBN 978-3-11-062939-2

25. Grüninger, M., Fox, M.S.: The role of competency questions in enterprise engineering. In: Rolstadås, A. (ed.) Benchmarking: Theory and Practice. IAICT, pp. 22–31. Springer, Boston (1995). https://doi.org/10.1007/978-0-387-34847-6_3. ISBN 978-0-412-62680-7

26. Hartl, N., Wössner, E., Sure-Vetter, Y.: Nationale Forschungsdateninfrastruktur (NFDI). Informatik Spektrum **44**(5), 370–373 (2021). https://doi.org/10.1007/s00287-021-01392-6

27. Herres-Pawlis, S.: NFDI4Chem: Fachkonsortium für die Chemie. Bausteine Forschungsdatenmanagement **2021**(2), 34–45 (2021). https://doi.org/10.17192/bfdm.2021.2.8340

28. Herres-Pawlis, S., Koepler, O., Steinbeck, C.: NFDI4Chem: shaping a digital and cultural change in chemistry. Angew. Chem. Int. Ed. **58**(32), 10766–10768 (2019). https://doi.org/10.1002/anie.201907260

29. Herres-Pawlis, S., Liermann, J.C., Koepler, O.: Research data in chemistry: results of the first NFDI4Chem community survey. Z. Anorg. Allg. Chem. **646**(21), 1748–1757 (2020). https://doi.org/10.1002/zaac.202000339

30. Höche, D., Konchakova, N., Zheludkevich, M., Hagelien, T., Friis, J.: Ontology assisted modelling of galvanic corrosion of magnesium. In: Chinesta, F., Abgrall, R., Allix, O., Kaliske, M. (eds.) Proceedings of WCCM-ECCOMAS 2020. Scipedia (2021). https://doi.org/10.23967/wccm-eccomas.2020.263

31. Horsch, M.T., et al.: Ontologies for the virtual materials marketplace. KI - Künstliche Intelligenz **34**(3), 423–428 (2020). https://doi.org/10.1007/s13218-020-00648-9

32. Horsch, M.T.: Mereosemiotics: parts and signs. In: Sanfilippo, E.M., et al. (eds.) Proceedings of JOWO 2021, p. 3. CEUR-WS (2021). http://ceur-ws.org/Vol-2969/

33. Horsch, M.T., Chiacchiera, S., Schembera, B., Seaton, M.A., Todorov, I.T.: Semantic interoperability based on the European materials and modelling ontology and its ontological paradigm: mereosemiotics. In: Chinesta, F., Abgrall, R., Allix, O., Kaliske, M. (eds.) Proceedings of WCCM-ECCOMAS 2020. Scipedia (2021). https://doi.org/10.23967/wccm-eccomas.2020.297

34. Horsch, M.T., et al.: Semantic interoperability and characterization of data provenance in computational molecular engineering. J. Chem. Eng. Data **65**(3), 1313–1329 (2020). https://doi.org/10.1021/acs.jced.9b00739

35. Horsch, M.T., Toti, D., Chiacchiera, S., Seaton, M.A., Goldbeck, G., Todorov, I.T.: OSMO: ontology for simulation, modelling, and optimization. In: Sanfilippo, E.M., et al. (eds.) Proceedings of JOWO 2021, p. 47. CEUR-WS (2021). http://ceur-ws.org/Vol-2969/

36. Jagusch, G.W., Preuß, N.: Umfragedaten zu "NFDI4Ing - Rückmeldung aus den Forschungscommunities". Data collection, NFDI4Ing (2019). https://doi.org/10.25534/tudatalib-104

37. Jupp, S., et al.: A new ontology lookup service at EMBL-EBI. In: Malone, J., Stevens, R., Forsberg, K., Splendiani, A. (eds.) Proceedings of SWAT4LS, pp. 118–119. CEUR-WS (2015). http://ceur-ws.org/Vol-1546/

38. Kchaou, M., Khlif, W., Gargouri, F., Mahfoudh, M.: Transformation of BPMN model into an OWL2 ontology. In: Ali, R., Kaindl, H., Maciaszek, L.A. (eds.) Proceedings of ENASE 2021, pp. 380–388. SciTePress (2021). ISBN 978-989-758-508-1

39. Klein, P., Preisig, H.A., Horsch, M.T., Konchakova, N.: Application of an ontology based process model construction tool for active protective coatings: corrosion inhibitor release. In: Sanfilippo, E.M., et al. (eds.) Proceedings of JOWO 2021, p. 26. CEUR-WS (2021). http://ceur-ws.org/Vol-2969/

40. Kraft, S., et al.: Nationale Forschungsdateninfrastruktur (NFDI) e.V.: Aufbau und Ziele. Bausteine Forschungsdatenmanagement **2021**(2), 1–9 (2021). https://doi.org/10.17192/bfdm.2021.2.8332
41. Liskin, O., Pham, R., Kiesling, S., Schneider, K.: Why we need a granularity concept for user stories. In: Cantone, G., Marchesi, M. (eds.) XP 2014. LNBIP, vol. 179, pp. 110–125. Springer, Cham (2014). https://doi.org/10.1007/978-3-319-06862-6_8. ISBN 978-3-319-06861-9
42. Metadata4Ing Working Group: Metadata4Ing: An ontology for describing the generation of research data within a scientific activity. version 1.0.0, NFDI4Ing (2022). https://nfdi4ing.pages.rwth-aachen.de/metadata4ing/metadata4ing/
43. Mittal, V.K., Bailin, S.C., Gonzalez, M.A., Meyer, D.E., Barrett, W.M., Smith, R.L.: Toward automated inventory modeling in life cycle assessment: the utility of semantic data modeling to predict real-world chemical production. ACS Sustain. Chem. Eng. **6**(2), 1961–1976 (2018). https://doi.org/10.1021/acssuschemeng.7b03379
44. NFDI-MatWerk: Die große Digitalisierungsumfrage. Data, NFDI-MatWerk (2020)
45. Ngandjong, A.C., et al.: Investigating electrode calendering and its impact on electrochemical performance by means of a new discrete element method model: towards a digital twin of Li-ion battery manufacturing. J. Power Sources **495**, 229320 (2021). https://doi.org/10.1016/j.jpowsour.2020.229320
46. Range, J., et al.: EnzymeML: a data exchange format for biocatalysis and enzymology. FEMS J. (2022). https://doi.org/10.1111/febs.16318
47. Riehl Figueiredo, L., Carvalho de Oliveira, H.: Automatic generation of ontologies from business process models. In: Hammoudi, S., Smialek, M., Camp, O., Filipe, J. (eds.) Proceedings of ICEIS 2018, pp. 81–91. SciTePress (2018). ISBN 978-989-758-298-1
48. Romanos, N., Kalogerini, M., Koumoulos, E.P., Morozinis, K., Sebastiani, M., Charitidis, C.: Innovative data management in advanced characterization: implications for materials design. Mater. Today Commun. **20**, 100541 (2019). https://doi.org/10.1016/j.mtcomm.2019.100541
49. Rospocher, M., Ghidini, C., Serafini, L.: An ontology for the business process modelling notation. In: Garbacz, P., Kutz, O. (eds.) Proceedings of FOIS 2014, pp. 133–146. IOS (2014). ISBN 978-1-61499-437-4
50. Rößler, M., Huth, P.U., Liauw, M.A.: Process analytical technology (PAT) as a versatile tool for real-time monitoring and kinetic evaluation of photocatalytic reactions. React. Chem. Eng. **5**(10), 1992–2002 (2020). https://doi.org/10.1039/d0re00256a
51. Schembera, B., Durán, J.M.: Dark data as the new challenge for big data science and the introduction of the scientific data officer. Philos. Technol. **33**(1), 93–115 (2019). https://doi.org/10.1007/s13347-019-00346-x
52. Schembera, B., Iglezakis, D.: EngMeta: metadata for computational engineering. Int. J. Metadata Semant. Ontol. **14**(1), 26–38 (2020). https://doi.org/10.1504/IJMSO.2020.107792
53. Schimmler, S., et al.: NFDI4Cat: local and overarching data infrastructures. In: Proceedings of e-Science Days 2021, heiBOOKS (2022, to appear)
54. Shao, Q., et al.: Material twin for composite material microstructure generation and reconstruction based on statistical continuum theory. Composites C **7**, 100216 (2022). https://doi.org/10.1016/j.jcomc.2021.100216
55. Strecker, D., Bossert, L.C., Demandt, É.: Das Versprechen der Vernetzung der NFDI. Bausteine Forschungsdatenmanagement **2021**(3), 39–55 (2021). https://doi.org/10.17192/bfdm.2021.3.8336

56. Theißen, M., Wiedau, M.: DEXPI P&ID specification. Version 1.3, ProcessNet, DEXPI Initiative (2021). https://dexpi.org/specifications/

57. Whetzel, P.L., et al.: BioPortal: enhanced functionality via new web services from the national center for Biomedical ontology to access and use ontologies in software applications. Nucleic Acids Res. **39**(S2), W541–W545 (2011). https://doi.org/10.1093/nar/gkr469

58. Wiesner, A., Morbach, J., Marquardt, W.: Information integration in chemical process engineering based on semantic technologies. Comput. Chem. Eng. **35**(4), 692–708 (2011). https://doi.org/10.1016/j.compchemeng.2010.12.003

59. Wiśniewski, D., Potoniec, J., Lawrynowicz, A., Keet, C.M.: Analysis of ontology competency questions and their formalizations in SPARQL-OWL. J. Web Semant. **59**, 100534 (2019). https://doi.org/10.1016/j.websem.2019.100534

60. Wulf, C., et al.: A unified research data infrastructure for catalysis research: challenges and concepts. ChemCatChem **13**(14), 3223–3236 (2021). https://doi.org/10.1002/cctc.202001974

Fast Predictions of Lattice Energies by Continuous Isometry Invariants of Crystal Structures

Jakob Ropers, Marco M. Mosca, Olga Anosova⬥, Vitaliy Kurlin$^{(\boxtimes)}$⬥, and Andrew I. Cooper

University of Liverpool, Liverpool L69 3BX, UK
`vkurlin@liv.ac.uk`
`http://kurlin.org`

Abstract. Crystal Structure Prediction (CSP) aims to discover solid crystalline materials by optimizing periodic arrangements of atoms, ions or molecules. CSP takes weeks of supercomputer time because of slow energy minimizations for millions of simulated crystals. The lattice energy is a key physical property, which hints at thermodynamic stability of a crystal but has no simple analytic expression. Past machine learning approaches to predict the lattice energy used slow crystal descriptors depending on manually chosen parameters. The new area of Periodic Geometry offers much faster isometry invariants that are also continuous under perturbations of atoms. Our experiments on simulated crystals confirm that a small distance between the new invariants guarantees a small difference of energies. We compare several kernel methods for invariant-based predictions of energy and achieve the mean absolute error of less than 5 kJ/mole or 0.05 eV/atom on a dataset of 5679 crystals.

Keywords: Crystal · Energy · Isometry invariant · Machine learning

1 Motivations, Problem Statement and Overview of Results

Solid crystalline materials (*crystals*) underpin key technological advances from solid-state batteries to therapeutic drugs. Crystals are still discovered by trial and error in a lab, because their properties are not yet expressed in terms of crystal geometries. This paper makes an important step towards understanding the structure-property relations, for example how an energy of a crystal depends on its geometric structure. The proposed methods belong to the recently established area of Periodic Geometry, which studies geometric descriptors (*continuous isometry invariants*) and metrics on a space of all periodic crystals.

The most important property of a crystal is the energy of its crystal structure, which is usually called the *lattice energy* or *potential energy surface* or *energy landscape* [22]. This lattice energy hints at thermodynamic stability of

© Springer Nature Switzerland AG 2022
A. Pozanenko et al. (Eds.): DAMDID/RCDL 2021, CCIS 1620, pp. 178–192, 2022.
https://doi.org/10.1007/978-3-031-12285-9_11

a crystal, whether such a crystal can be accessible for synthesis in a lab and can remain stable under application conditions. Since the lattice energy has no closed analytic expression, calculations are always approximate, from the *force field* (FF) level [12] to the more exact density functional theory (DFT) [8].

Our experiments use the lattice energy obtained by force fields for the CSP data of 5679 nanoporous T2 crystals in Fig. 1 predicted by our colleagues [16].

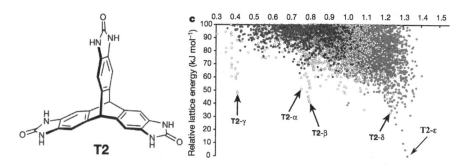

Fig. 1. Left: T2 molecule. **Right:** energy-vs-density plot of 5679 predicted crystals, five polymorphs were synthesized, most recent T2-ε crystal is added to [16, Fig. 2d].

Traditionally a periodic crystal is stored in a Crystallographic Information File (CIF). This file specifies a linear basis v_1, v_2, v_3 of $\mathbb{R}^3$, which spans the *unit cell* $U = \{\sum_{i=1}^{3} c_i v_i \mid 0 \leq c_i < 1\}$, generates the *lattice* $\Lambda = \{\sum_{i=1}^{3} c_i v_i \mid c_i \in \mathbb{Z}\}$. Then a crystal can be obtained as the infinite union of lattice translates $M + \Lambda = \{p + v : p \in M, v \in \Lambda\}$ from a finite set (*motif*) of points $M \subset U$ in the cell U. The representation $M + \Lambda$ is simple but is highly ambiguous in the sense that infinitely many pairs (cell, motif) generate equivalent crystals, see [2, Fig. 2].

The main novelty of our approach to energy predictions is using the fast computable and easily interpretable invariants of crystals. The concept of an invariant has a rigorous definition after we fix an equivalence relation on objects in question. Since crystal structures are determined in a rigid form, the most natural equivalence is rigid motion or *isometry*, which is any map that preserves interpoint distances, for example a composition of translations and rotations in Euclidean space $\mathbb{R}^3$. Any orientation-preserving isometry can be realized as a *rigid motion*, which is a continuous family f_t, $t \in [0, 1]$, of isometries starting from the identity map $f_0 = \mathrm{id}$. Since any general isometry is a composition of a single reflection and a rigid motion, we consider isometry as our main *equivalence relation* on crystals. Later we can also take into account a sign of orientation.

An *isometry invariant* I is a crystal property or a function, say from crystals to numbers, preserved by isometry. So if crystals S, Q are isometric then $I(S) = I(Q)$. The classical example invariants of a crystal S are the symmetry group (the group of isometries that map S to itself) and the volume of a minimal (*primitive*) unit cell. Example non-invariants are unit cell parameters (edge-lengths and angles) and fractional coordinates of atoms in a cell basis.

Many widely used isometry invariants including symmetry groups break down (are *discontinuous*) under perturbations of atoms, which always exist in real crystals at finite temperature. Perturbations are also important for distinguishing simulated crystals obtained via Crystal Structure Prediction (CSP). Indeed, CSP iteratively minimizes the lattice energy and inevitably stops at some approximation to a local minimum [13]. Hence, after many random initializations, we likely get many near duplicate structures around the same local minumum.

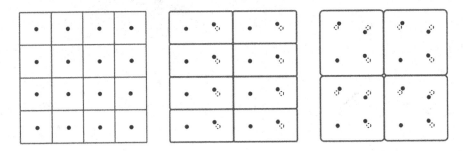

Fig. 2. Most past invariants are discontinuous under perturbations above, for example symmetry groups and sizes of primitive or reduced cells. Recent isometry invariants [10,24,25] continuously quantify similarities between perturbed periodic structures.

Since any perturbation of points keeping their periodicity (but not necessarily an original unit cell) produces a new close structure, all periodic structures form a continuous space. Then any CSP dataset can be viewed as a discrete sample from the underlying continuous space of periodic structures. The lattice energy is a function on this crystal space whose geometry needs to be understood. The problem below is a key step towards describing structure-property relations.

Properties-from-Invariants Problem. Find suitable isometry invariants that justifiably predict desired properties of crystals such as the lattice energy.   ∎

The proposed invariants to tackle the above problem are average minimum distances (AMD) [25]. AMD is an infinite sequence of isometry invariants whose values change by at most 2ε if given points are perturbed in their ε-neighborhoods. A thousand of AMD invariants can be computed in milliseconds on a modest desktop for crystals with hundreds of atoms in a unit cell [25, Appendix D].

The above continuity of AMD guarantees that perturbed crystals have close AMD values. Then such a theoretically continuous invariant can be tested for checking continuity of energy under crystal perturbations. The first contribution is an experimental detection of constants λ and δ such that, for any smaller distance $d < \delta$ between AMD vectors, the corresponding crystals have a lattice energy difference within λd, usually within 2kJ/mole. Past invariants have no such a constant to quantify continuity of energy in this way. For example, close values of density, RMSD [4], PXRD [18] don't guarantee close values of energy.

The second contribution is the demonstration that several kernel methods can achieve a mean absolute error of less than 5 kJ/mole by using only isometry invariants without any chemical data. The key achievement is the time of less than 10 min for training by using a modest desktop on a dataset of 5679 structures, while energy predictions take milliseconds per crystal on average.

Section 2 reviews closely related past work using crystal descriptors for machine learning of the lattice energy. Section 3 reminds the recently introduced isometry invariants of periodic point sets and their properties. Section 4 quantifies continuity of energy in terms of AMD invariants. Section 5 describes how the energy of a crystal can be predicted from its AMD invariants by using several kernel methods. Section 6 discusses limitations and potential developments.

2   Review of Related Machine Learning Approaches

This section reviews the closest related work about energy predictions for infinite periodic crystals. The same problem is simpler for a single molecule [20].

Energy predictions use various representations of crystals. We review only geometric descriptors that are closest to isometry invariants in Sect. 3.

The partial radial distribution function (PRDF) is based on the density of atoms of type β in a shell of radius r and width dr centered around an atom of type α [19]. Since atom types are essentially used, the PRDF can be best for comparing crystals that are composed of the same atom types. Due to averaging across all atoms of a type α within a unit cell, the PRDF is independent of a cell choice. A similar distance-based fingerprint was introduced earlier by Valle and Oganov [21]. Since only pairwise distances are used, these descriptors are isometry invariants and likely continuous under perturbations shown in Fig. 2.

Completeness or uniqueness of a crystal with a given PRDF is unclear yet, but can be theoretically possible for a large enough radius r. Practical computations require choices or the distance thresholds r and dr, which can affect the PRDF. Schutt et al. confirm in [19, Table I] that the PRDF outperforms non-invariant features such as the Bravais matrix of cell parameters. The mean absolute error (MAE) of energy predictions based on PRDF is 0.68 eV/atom or 65.6 kJ/mole.

Another way to build geometric attributes of a crystal structure is to use Wigner-Seitz cells (also called Dirichlet or Voronoi domains) of atoms. Ward et al. [23] used 271 cell-based geometric and chemical attributes to achieve the MAE of 0.09 eV/atom or 8.7 kJ/mole for predicting the formation enthalpy.

An extensible neural network potential [20, Fig. 4] has further improved the mean absolute error (MAE) to 1.8 kcal/mole = 7.56 kJ/mole. The most advanced approach by Egorova et al. [6] predicts the difference between the accurate DFT energy and its force field approximation a with MAE less than 2 kJ/mole by using GGA DFT (PBE) calculations and symmetry function descriptors [3].

3   Key Definitions and Recent Results of Periodic Geometry

This section reviews more recent work in the new area of Periodic Geometry [1], which studies the metric geometry on the space of all periodic structures. Nuclei of atoms are better defined physical objects than chemical bonds, which depend on many thresholds for distances and angles. Hence the most fundamental model of a crystal is a periodic set of zero-sized points representing all atomic centers.

Though chemical elements and other physical properties can be easily added to invariants as labels of points, the experiments in [5,25] and Sects. 4, 5 show that the new invariants can be enough to infer all chemistry from geometry.

The symbol $\mathbb{R}^n$ denotes Euclidean space with *Euclidean* distance $|p - q|$ between points $p, q \in \mathbb{R}^n$. Motivated by a traditional representation of a crystal by a Crystallographic Information File, a periodic point set S is given by a pair (cell U, motif M). Here U is a *unit cell* (parallelepiped) spanned by a linear basis $v_1, \ldots, v_n$ of $\mathbb{R}^n$, which generates the *lattice* $\Lambda = \{\sum_{i=1}^{n} c_i v_i : c_i \in \mathbb{Z}\}$. A *periodic point set* $S = M + \Lambda$ is obtained by shifting a finite *motif* $M \subset U$ of points along all vectors $v \in \Lambda$. Figure 3 illustrates the problem of transforming ambiguous input into invariants that can distinguish periodic sets up to isometry.

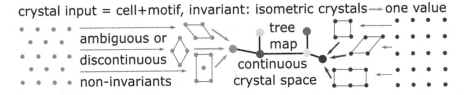

crystal input = cell+motif, invariant: isometric crystals → one value

ambiguous or
discontinuous
non-invariants

tree
map
continuous
crystal space

Fig. 3. Any periodic sets, for example the hexagonal and square lattices, can be represented by infinitely many pairs (cell, motif). This ambiguity can be resolved only by a complete isometry invariant that should continuously parameterize the crystal space.

Arguably the simplest isometry invariant of a crystal is its density ρ. Without distinguishing atoms in a periodic point set S, the density $\rho(S)$ is the number m of points in a unit cell U, divided by the cell volume Vol$[U]$. The density ρ distinguishes hexagonal and square lattices in Fig. 3 but is insensitive to perturbations shown in Fig. 2. Though many real crystals are dense and can not be well-separated by density, energy landscapes are still visualized as energy-vs-density plots in Fig. 1. The single-value density ρ has been recently extended to the sequence of density functions $\psi_k(t)$ [5]. For any integer $k \geq 1$, the *density function* $\psi_k(t)$ measures the volume of the regions within a unit cell U covered by k balls with radius $t \geq 0$ and centers at all points $p \in M$, divided by Vol$[U]$.

Though these isometry invariants have helped to identify a missing crystal in the Cambridge Structural Database, their running time cubically depends on k, which is a bit slow for big datasets. The following invariants are much faster.

Let a periodic point set $S \subset \mathbb{R}^n$ have points $p_1, \ldots, p_m$ in a unit cell. For any $k \geq 1$ and $i = 1, \ldots, m$, the i-th row of the $m \times k$ matrix $D(S; k)$ consists of the ordered distances $d_{i1} \leq \cdots \leq d_{ik}$ measured from the point p_i to its first k nearest neighbors within the infinite set S, see Fig. 4. The *Average Minimum Distance* $\mathrm{AMD}_k(S) = \dfrac{1}{m} \sum\limits_{i=1}^{m} d_{ik}$ is the average of the k-th column in $D(S; k)$.

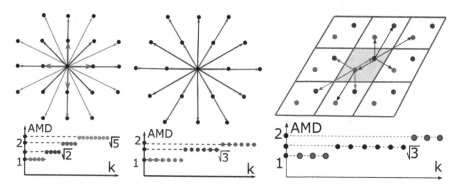

Fig. 4. [25, Fig. 4] **Left:** in the square lattice, distances from the origin to its first few neighbors are shown in the graph of AMD_k values, e.g. the shortest axis-aligned distances are $\mathrm{AMD}_1 = \cdots = \mathrm{AMD}_4 = 1$, the longer diagonal distances are $\mathrm{AMD}_5 = \cdots = \mathrm{AMD}_8 = \sqrt{2}$. **Middle:** in the hexagonal lattice, the shortest distances are $\mathrm{AMD}_1 = \cdots = \mathrm{AMD}_6 = 1$. **Right:** AMD for a honeycomb periodic set (graphene).

[25, Theorem 4] proves that AMD is an isometry invariant independent of a unit cell. The AMD invariants are similar to radial distribution functions [19] and related density-based invariants [21]. The AMD definition has no manually chosen thresholds such as cut-off radii or tolerances. The length k of the vector $\mathrm{AMD}^{(k)} = (\mathrm{AMD}_1, \ldots, \mathrm{AMD}_k)$ is not a parameter in the sense that increasing k only adds new values without changing previous ones. Hence k can be considered as an order of approximation, similarly to an initial length of a DNA code.

We have no examples of non-isometric sets that have identical infinite AMD sequences. Hence AMD can be complete at least for periodic sets in general position so that if two sets S, Q have $\mathrm{AMD}(S) = \mathrm{AMD}(T)$, then S, Q are isometric. More recently, the isometry classification of all periodic point sets was reduced to an *isoset* [2], which is a collection of atomic environments considered modulo rotations and up to a *stable* radius α. This stable radius is defined for a given crystal and any two crystals can be compared by isosets of their maximum radius so that two sets S, Q are isometric if and only if their isosets are equivalent.

This paper uses AMD invariants due to their easy interpretability and fast running time. $\mathrm{AMD}_k(S)$ asymptotically approaches $c(S) \sqrt[n]{k}$, where $c(S)$ is related to the density of a periodic point set $S \subset \mathbb{R}^n$, see [25, Theorem 13]. A near linear computational time [25, Theorem 14] of AMD_k in both m, k translates into milliseconds on a modest laptop, which allowed us to visualize all 229K organic molecular crystals from the Cambridge Structural Database in a few hours.

4   Continuity of the Energy in Terms of AMD Invariants

To express continuity of AMD and other invariants under perturbations, we use the maximum displacement of atoms formalized by the *bottleneck distance* d_B as follows. For any bijection $g : S \to Q$ between periodic point sets, the maximum displacement is $d_g(S, Q) = \sup_{p \in S} |g(p) - p|$. After minimizing over all bijections $g : S \to Q$, we get the *bottleneck distance* $d_B(S, Q) = \inf_{g:S \to Q} d_g(S, Q)$.

The structure-property hypothesis says that all properties of a crystal should be determined by its geometric structure. Understanding how any property can be explicitly computed from a crystal structure would replace trial-and-error methods by a guided discovery to find crystals with desired properties.

Most current attempts are based on black-box machine learning of properties from crystal descriptors, not all of which are invariants up to isometry. All machine learning tools rely on the usually implicit assumption that small perturbations in input data lead to relatively small perturbations in outputs.

Continuity of a structure-property relation can be mathematically expressed as Lipschitz continuity [14, Sect. 9.4]: $|E(S) - E(Q)| \leq \lambda d(S, Q)$, where λ is a constant, E is a crystal property such as the lattice energy, $d(S, Q)$ is a distance satisfying all metric axioms on crystals S, Q or their invariants. The above inequality should hold for all crystals S, Q with small distances $d(S, Q) < \delta$, where a threshold δ may depend on a property E or a metric d, not on S, Q.

The continuity above sounds plausible and seems necessary for the structure-property hypothesis. Indeed, if even small perturbations of a geometric structure drastically change crystal properties, then any inevitably noisy structure determination would not suffice to guarantee desired properties of a crystal.

Figure 5, 6, 7 show that the past methods of characterizing crystal similarity are insufficient to guarantee the above continuity of the lattice energy. These results were obtained on the T2 dataset of 5679 simulated crystals reported in [16]. Each square dot represents a pair of crystals with differences in past descriptors on the horizontal axis and differences in energies on the vertical axis.

Figure 5 shows dozens of crystal pairs with very close densities and rather different lattice energies, which means that the energy discontinuously varies relative to the density. This failure of a single-value descriptors might not be surprising not only for crystals, which are often very dense materials, but also for other real-life scenarios. For example, many people have the same height and very different weights. However, the density is still used to represent a crystal structure in CSP landscapes such as Fig. 1. Indeed, the density is an isometry invariant, which is continuous (actually, constant) under perturbations, see Fig. 2.

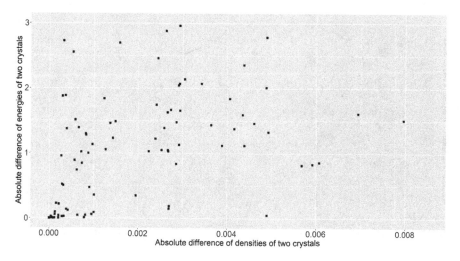

Fig. 5. 5679 crystals in Fig. 1 have the density range [0.3, 1.4]. Many crystals have differences in densities within 0.003 g/cm³ and differences in energies up to 3 kJ/mole.

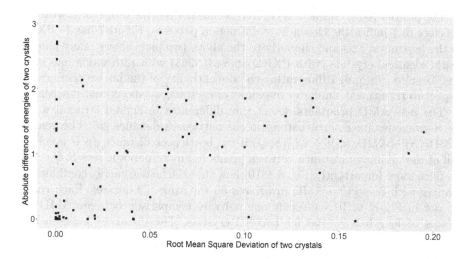

Fig. 6. Crystal pairs with RMSD < 0.1Å have differences in energies up to 3 kJ/mole.

Figure 6 illustrates a similar conclusion for the traditional packing similarity measured by the COMPACK algorithm [4] as the Root Mean Square Deviation (RMSD) of atomic positions matched between up to 15 (by default) molecules in two crystals. This similarity relies on two extra thresholds for atomic distances and angles whose values affect the RMSD. For example, when only one of 15 molecules is matched, the RMSD is exactly 0, because all 5679 crystals are based on the same T2 molecule in Fig. 1. Nonetheless, this packing similarity can visually confirm that nearly identical crystals nicely overlap each other.

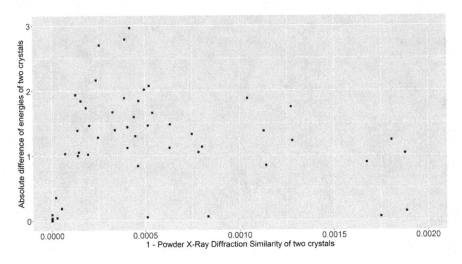

Fig. 7. Crystal pairs with PXRD similarity > 0.9995 have big differences in energies.

The powder X-ray diffraction (PXRD) similarity has the range $[0, 1]$ with values close to 1 indicating closeness of diffraction patterns. Figure 7 has 1−PXRD on the horizontal axis and similarly to the above two plots shows many pairs of nearly identical crystals (with PXRD above 0.9995) with rather different energies. Despite Fig. 5, 6, 7 illustrating the discontinuity of the lattice energy with respect to traditional similarity measures of crystals, we should not despair.

The new AMD invariants detect tiny differences in crystal structures and are continuous under perturbations in the bottleneck distance [25, Theorem 9]: $|\text{AMD}_k(S) - \text{AMD}_k(Q)| \leq 2d_B(S, Q)$ if the bottleneck distance d_B is less than half of the minimum distance between points in any of periodic sets $S, Q \subset \mathbb{R}^n$.

Even more importantly, Fig. 8, 9,10 show that the lattice energy continuously changes with respect to AMD invariants on the same T2 dataset. Each rhombic dot in Fig. 8, 9, 10 represents one pairwise comparison between $\text{AMD}^{(100)}$ vectors of length $k = 100$ for two T2 crystals. The distances between vectors $p = (p_1, \ldots, p_k)$ and $q = (q_1, \ldots, q_k)$ on the horizontal axis are computed by the Euclidean metric $L_2(p, q) = \sqrt{\sum_{i=1}^{k} |p_i - q_i|^2}$, the Chebyshev metric $L_\infty(p, q) = \max_{i=1,\ldots,k} |p_i - q_i|$ and the Manhattan metric $L_1(p, q) = \sum_{i=1}^{k} |p_i - q_i|$.

Despite the T2 dataset being thoroughly filtered out to remove near duplicates, Fig. 8, 9 include several pairs whose AMD invariants are very close, though not identical. In all these cases the corresponding crystals also have very close energies, which can be quantified via Lipschitz continuity as follows.

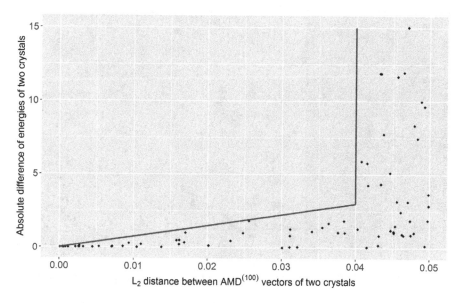

Fig. 8. The green line $|\Delta E| = 75L_2$ over $L_2 \in [0, 0.04]$ shows that if crystals have a distance $L_2 < 0.04\text{\AA}$ between $\text{AMD}^{(100)}$ vectors, their energies differ by at most $75L_2$. (Color figure online)

In Fig. 8 the Lipschitz continuity for the energy $|\Delta E| = |E(S) - E(Q)| \leq \lambda_2 L_2(\text{AMD}^{(100)}(S), \text{AMD}^{(100)}(Q))$ holds for $\lambda_2 = 200$ and all pairs of crystals S, Q whose $\text{AMD}^{(100)}$ vectors have a Euclidean distance $L_2 < \delta_2 = 0.04\text{\AA}$.

Visually, all these pairs are below the green line $\Delta E = 200L_2$ up to the distance threshold $\delta_2 = 0.04\text{\AA}$. If distances between crystals become too large, a single-value metric cannot guarantee close values of energy. Using the geographic analogy, the further we travel from any fixed location on planet Earth, the more variation in physical properties such as the altitude we should expect.

Figure 9 illustrates continuity of the lattice energy with respect to the metric $L_\infty(p, q) = \max\limits_{i=1,\dots,k} |p_i - q_i|$ between $\text{AMD}^{(100)}$ vectors. All pairs of crystals with distances $L_\infty < \delta_\infty = 0.009\text{\AA}$ have differences in energies less than $\lambda_\infty L_\infty$ with $\lambda_\infty = 200$, so all corresponding dots are below the green line $|\Delta E| = 200L_\infty$.

Figure 10 shows that the lattice energy continuously behaves for the metric $L_1(p, q) = \sum\limits_{i=1}^{k} |p_i - q_i|$ between $\text{AMD}^{(100)}$ vectors. All pairs of crystals with distances $L_1 < \delta_1 = 0.32\text{\AA}$ have energy differences less than $\lambda_1 L_1$ with $\lambda_1 = 10$, so all corresponding dots are below the green line $|\Delta E| = 10L_1$.

The thresholds $\delta_1 = 0.32$ and $\delta_2 = 0.04$ are larger than $\delta_\infty = 0.009\text{\AA}$, because the metrics L_1, L_2 sum up all deviations between corresponding coordinates of $\text{AMD}^{(100)}$ vectors, while the metric L_∞ measures only the maximum deviation.

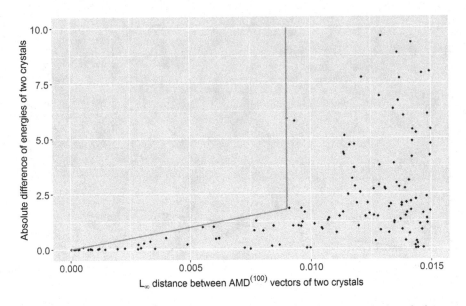

Fig. 9. The green line $|\Delta E| = 200L_\infty$ over $[0, 0.009]$ shows that if crystals have a distance $L_\infty < 0.009\text{Å}$ between $\text{AMD}^{(100)}$, their energies differ by at most $200L_\infty$. (Color figure online)

If we tried to fit Lipschitz continuity for the past descriptors (density, RMSD, PXRD) in Fig. 5, 6, 7 similarly to AMD invariants above, corresponding green lines would be almost vertical with huge slopes or gradients (Lipschitz constants).

5   Fast Predictions of the Energy by AMD Invariants

This section describes the second important contribution by showing that continuity of AMD from Sect. 4 leads to state-of-the-art energy predictions.

The energy prediction problem is to infer the lattice energy from a crystal structure, for example by using a dataset of ground truth energies for training.

The descriptors in Fig. 5, 6, 7 cannot be justifiably used to resolve the above problem because of their discontinuity. Indeed, if we input a slightly different (say, experimental) crystal, we expect a close value of energy in the output.

First we describe the Gaussian Process Regression [9] as implemented in SciKit Learn [15], see Fig. 11, which achieved the best results on the T2 dataset of 5679 crystals. Initially each T2 crystal is converted into a periodic point set S by placing a zero-sized point at every atomic center. Then each S is represented by its $\text{AMD}^{(k)}(S)$ vector of a fixed length in the range $k = 50, 100, \ldots, 500$. The base distance d between $\text{AMD}^{(k)}$ vectors was chosen as L_∞ due to the smallest Lipschitz constant $\lambda = 2$ in the continuity property $|\text{AMD}_k(S) - \text{AMD}_k(Q)| \leq \lambda d_B(S, Q)$. For the metrics L_1, L_2, the Lipschitz constants would be $2k, 2\sqrt{k}$.

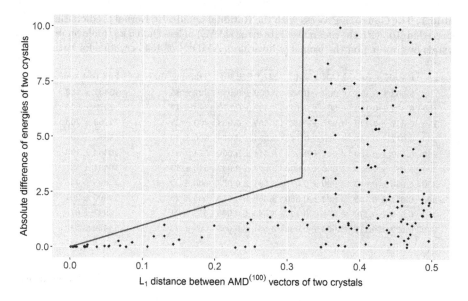

Fig. 10. The green line $|\Delta E| = 10L_1$ over $L_1 \in [0, 0.32]$ shows that if crystals have a distance $L_1 < 0.32\text{Å}$ between $\text{AMD}^{(100)}$ vectors, their energies differ by at most $10L_1$. (Color figure online)

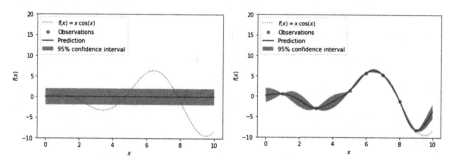

Fig. 11. Gaussian Process tries to predict values of $f(x) = x\cos x$ by training on observed data points. **Left:** an initial prediction is 0 for any x. **Right:** predictions substantially improve after training on six data points under the natural assumption that the underlying function is *continuous*, so continuity is important for learning.

For any pair of crystals S, Q, we consider the Rational Quadratic Kernel $K(S, Q) = \left(1 + \dfrac{d^2(S, Q)}{2\alpha l^2}\right)^{-\alpha}$, where α, l are scale parameters optimized by training. For a single prediction run, the whole T2 dataset was randomly split into 80% training subset and remaining 20% test subset of $m = 1136$ crystals.

Table 1. The Gaussian Process with the Rational Quadratic Kernel predicts the energy reported in [16] with the mean absolute error (MAE) of less than 5 kJ/mole on $m = 1136$ crystals by training on the isometry invariants $AMD^{(k)}$ of 4543 crystals for various k.

k	RMSE ± std	MAE ± std	MAPE ± std	Training time, sec	Full test time, ms
50	6.503 ± 0.123	4.900 ± 0.86	3.509 ± 0.059	627 ± 85	15961 ± 183
100	6.344 ± 0.152	4.801 ± 0.103	3.439 ± 0.070	349 ± 47	7979 ± 564
150	6.607 ± 0.119	4.977 ± 0.077	3.559 ± 0.053	400 ± 23	12789 ± 203
200	6.617 ± 0.147	4.966 ± 0.114	3.554 ± 0.079	506 ± 40	15943 ± 46
250	6.517 ± 0.109	4.914 ± 0.082	3.514 ± 0.055	574 ± 91	16464 ± 193
300	6.632 ± 0.139	5.003 ± 0.092	3.577 ± 0.062	545 ± 15	16431 ± 52
350	6.615 ± 0.077	4.990 ± 0.077	3.581 ± 0.053	500 ± 22	12395 ± 44
400	6.611 ± 0.149	4.984 ± 0.080	3.569 ± 0.053	585 ± 25	17906 ± 201
450	6.559 ± 0.179	4.954 ± 0.127	3.545 ± 0.085	512 ± 21	12927 ± 67
500	6.622 ± 0.116	5.004 ± 0.092	3.581 ± 0.068	598 ± 24	18429 ± 219

Table 1 shows three types of errors, each averaged over 10 runs above: $RMSE = \sqrt{\dfrac{1}{m} \sum_{i=1}^{m} |E_{true}(S_i) - E_{pred}(S_i)|^2}$ is the root mean square error in the lattice energy averaged over m crystals $S_1, \ldots, S_m$ from the test subset, then $MAE = \dfrac{1}{m} \max_{i=1,\ldots,m} |E_{true}(S_i) - E_{pred}(S_i)|$ is the mean absolute error and the mean absolute percentage error $MAPE = \dfrac{1}{m} \max_{i=1,\ldots,m} \dfrac{|E_{true}(S_i) - E_{pred}(S_i)|}{E_{true}(S_i)}$. Each value has the empirical standard deviation ±std computed over 10 runs.

Table 1 shows that the errors RMSE, MAE, MAPE are consistent across different values of k. The key advantage over past methods is the speed: less than 10 min for training for 4543 vectors $AMD^{(k)}$ on Intel Xeon CPU at 2.3 GHz. The last column shows the full test time on $m = 1136$ crystals, so the average time per crystal is more than 1000 times faster. The smallest mean absolute error $MAE \approx 4.8$ kJ/mole corresponds to about 7.4 (ms) per crystal.

The computation of $AMD^{(k)}$ asymptotically has a near linear time in k and the number of atoms in a unit cell by [25, Theorem 14], which needs only 27 ms on average per T2 crystal for $k = 1000$ on a similar desktop. This ultra-fast speed allowed us to visualize for the first time all 229K molecular organic crystals from the Cambridge Structural Database in less than 9 h, see [25, Appendix D].

We have tried other types of kernels: the matern and linear kernels gave slightly larger errors, the squared exponential was worse for some k. We also considered another version of the T2 dataset without hydrogens (32 atoms per molecule instead of 46), which gave a bit bigger error for all kernels above.

Instead of AMD invariants, we trained the Gaussian Process Regression on the density functions $\psi_k(t)$ [5], which are continuous isometry invariants extending the single-value density for a variable radius $t \geq 0$. The average errors of AMD-based predictions were smaller than for the density functions ψ_k, which are also slower to compute than AMD, asymptotically in a cubic time in k.

Finally, the Random Forest [11] and Dense Neural Network [7] trained on AMD and density functions performed slight worse than the Gaussian Process, though the training and text times were much faster (seconds instead of minutes). The experiments above are reported in the dissertation of the first author [17].

6   Conclusions and a Discussion of Future Developments

This paper has demonstrated that the recently developed continuous isometry invariants can provide insights undetected by traditional similarity measures.

In Sect. 4 Fig. 5, 6, 7 show that many crystals can have almost identical density, RMSD, PXRD patterns but rather different lattice energies. On the same T2 dataset [16] Fig. 8, 9, 10 show that the lattice energy satisfies the Lipschitz continuity $|E(S) - E(Q)| \leq \lambda d(S,Q)$ for a fixed constant λ and all crystals S, Q whose AMD invariants are close with respect to the metrics L_1, L_2, L_∞.

In Sect. 5 the standard kernel methods trained only on 100 isometry invariants $\text{AMD}^{(100)}$ achieved the state-of-the-art mean absolute error of less than 5 kJ/mole in energy. The key achievement is the speed of training (about 10 min for 4543 crystals on a modest desktop) and testing, which run in milliseconds per crystal. The code of experiments in Sect. 5 is available on GitHub [17].

It should not be surprising that the lattice energy can be efficiently predicted from distance-based invariants without any chemical information. Indeed, if one atom is replaced by a different chemical element, then inter-atomic distances to neighbors inevitably change, even if slightly. These differences in distances can be detected, also after averaging over all motif points. So AMD invariants should pick up differences in crystals after swapping chemically different atoms.

We don't know any non-isometric periodic point sets that have identical infinite sequences $\{\text{AMD}_k\}_{k=1}^{+\infty}$. Claiming such a counter-example to completeness requires a theoretical proof, because any computation outputs AMD values only up to a finite k. The recent work [24] studies the stronger Pointwise Distance Distributions, which avoid averaging over motif points, but keep the isometry invariance and continuity under perturbations of points for a suitable metric.

Acknowledgements. Supported by £3.5M EPSRC grant 'Application-driven Topological Data Analysis'.

References

1. Anosova, O., Kurlin, V.: Introduction to periodic geometry and topology. arXiv:2103.02749 (2021)
2. Anosova, O., Kurlin, V.: An isometry classification of periodic point sets. In: Lindblad, J., Malmberg, F., Sladoje, N. (eds.) DGMM 2021. LNCS, vol. 12708, pp. 229–241. Springer, Cham (2021). https://doi.org/10.1007/978-3-030-76657-3_16
3. Behler, J.: Atom-centered symmetry functions for constructing high-dimensional neural network potentials. J. Chem. Phys. **134**(7), 074106 (2011)
4. Chisholm, J., Motherwell, S.: COMPACK: a program for identifying crystal structure similarity using distances. J. Appl. Crystallogr. **38**(1), 228–231 (2005)

5. Edelsbrunner, H., Heiss, T., Kurlin, V., Smith, P., Wintraecken, M.: The density fingerprint of a periodic point set. In: Proceedings of SoCG (2021)
6. Egorova, O., Hafizi, R., Woods, D.C., Day, G.M.: Multifidelity statistical machine learning for molecular crystal structure prediction. J. Phys. Chem. A **124**(39), 8065–8078 (2020)
7. Goodfellow, I., Bengio, Y., Courville, A.: Deep Learning, vol. 1. MIT Press, Cambridge (2016)
8. Gross, E., Dreizler, R.: Density Functional Theory, vol. 337. Springer, Heidelberg (2013)
9. KI Williams, C.: Gaussian Processes for Machine Learning. Taylor & Francis (2006)
10. Mosca, M., Kurlin, V.: Voronoi-based similarity distances between arbitrary crystal lattices. Cryst. Res. Technol. **55**(5), 1900197 (2020)
11. Myles, A.J., Feudale, R.N., Liu, Y., Woody, N.A., Brown, S.D.: An introduction to decision tree modeling. J. Chemom. **18**(6), 275–285 (2004)
12. Niketic, S.R., Rasmussen, K.: The Consistent Force Field: A Documentation, vol. 3. Springer, Heidelberg (2012)
13. Oganov, A.: Modern Methods of Crystal Structure Prediction. Wiley, Hoboken (2011)
14. O'Searcoid, M.: Metric Spaces. Springer, Heidelberg (2006). https://doi.org/10.1007/978-1-84628-627-8
15. Pedregosa, F., et al.: Scikit-learn: machine learning in python. J. Mach. Learn. Res. **12**, 2825–2830 (2011)
16. Pulido, A., et al.: Functional materials discovery using energy-structure maps. Nature **543**, 657–664 (2017)
17. Ropers, J.: Applying machine learning to geometric invariants of crystals (2021). https://github.com/JRopes/CrystalEnergyPrediction
18. Sacchi, P., Lusi, M., Cruz-Cabeza, A.J., Nauha, E., Bernstein, J.: Same or different-that is the question: identification of crystal forms from crystal structure data. CrystEngComm **22**(43), 7170–7185 (2020)
19. Schütt, K., Glawe, H., Brockherde, F., Sanna, A., Müller, K.R., Gross, E.: How to represent crystal structures for machine learning: towards fast prediction of electronic properties. Phys. Rev. B **89**(20), 205118 (2014)
20. Smith, J., Isayev, O., Roitberg, A.: An extensible neural network potential with DFT accuracy at force field computational cost. Chem. Sci. **8**, 3192–3203 (2017)
21. Valle, M., Oganov, A.R.: Crystal fingerprint space-a novel paradigm for studying crystal-structure sets. Acta Crystallogr. A **66**(5), 507–517 (2010)
22. Wales, D.J.: Exploring energy landscapes. Annu. Rev. Phys. Chem. **69**, 401–425 (2018)
23. Ward, L., et al.: Including crystal structure attributes in machine learning models of formation energies via voronoi tessellations. Phys. Rev. B **96**(2), 024104 (2017)
24. Widdowson, D., Kurlin, V.: Pointwise distance distributions of periodic sets. https://arxiv.org/abs/2108.04798
25. Widdowson, D., Mosca, M., Pulido, A., Kurlin, V., Cooper, A.: Average minimum distances of periodic point sets—foundational invariants for mapping all periodic crystals. MATCH Commun. Math. Comput. Chem. **87**(3), 529–559 (2022)

Image Recognition for Large Soil Maps Archive Overview: Metadata Extraction and Georeferencing Tool Development

Nadezda A. Vasilyeva[1]([✉])(iD), Artem Vladimirov[1,2](iD), and Taras Vasiliev[1](iD)

[1] Dokuchaev Soil Science Institute, Pyzhevsky per. 7/2,
119017 Moscow, Russian Federation
nadezda.vasilyeva@gmail.com

[2] Bogoliubov Laboratory of Theoretical Physics, Joint Institute for Nuclear Research, Joliot-Curie 6, 141980 Dubna, Russian Federation

Abstract. During the second half of 20th century Dokuchaev Soil Science Institute has collected soil maps for a half of the Eurasian continent as a result of large national soil surveys which lasted for several decades with the efforts of the former USSR. Such labor-intensive expeditions on countries scale were not repeated since then. The question of future soil dynamics as Earth's fertile layer became crucial with global population growth and causes large part of uncertainty in Earth System Modelling. Most of the present knowledge about soil types is still in form of paper soil maps, representing valuable knowledge about soil cover of the past. Soil type itself is a crucial factor which still cannot be determined remotely but can be updated. Archive soil maps (several thousands of sheets) are an example of data which require digitizing and could profit from application of image recognition techniques. In the current study we present a demo tool for fast extraction of metadata and geo-referencing of paper soil maps using image recognition techniques. Presented software can be used for creating soil maps digital catalog allowing for a quick overview of a large collection.

Keywords: Soil maps · Archive collection · Image recognition · Heterogeneous data · Reproducibility

1 Introduction

Rapid progress in measuring instruments make soil studies a data-intensive domain with different spatio-temporal scales. Massive soil data come from field observations (i.e. continuous sensor measurements in precision agriculture), large raw data come from laboratory analysis such as spectrometry, mass spectrometry, microscopy, tomography, and analysis of proximal and remote sensing data and images (from hand equipment to drones and satellites) as well as scanned soil maps. Soil research historically developed as data-driven thus its progress revolutionized with modern techniques [16].

© Springer Nature Switzerland AG 2022
A. Pozanenko et al. (Eds.): DAMDID/RCDL 2021, CCIS 1620, pp. 193–204, 2022.
https://doi.org/10.1007/978-3-031-12285-9_12

An example of global spatial soil information is a soil type depicted in soil maps, which is a crucial factor for estimating suitability of land for different use, civil engineering, related natural resource planning and strategies for soil protection and sustainable land use. Variety of soil types are shown on soil maps of different scale and in different level of detalization. Soil maps are the results of large scale soil surveys based on field expeditions all over the country, involving enormous number of soil experts, thus it is an expensive and labor-consuming task which can not be repeated even every century for such countries as Russia. Rather update of existing soil maps using modern remote techniques and data analysis could be efficient. Archive map documents are meant those originally published on paper before switch to digital map production. Most of soil maps are still stored in traditional libraries and are retrievable just by bibliographic information. Additionally, many of these documents are over 50 years old and are beginning to deteriorate [14]. There two main stages/goals in map processing: 1) preservation of existing paper maps in a digital format and preventing them from loss or deterioration; 2) digitisation of all of its spatial information for use in a combined analysis with other space-time data.

2   Related Work

Recently most libraries around the world started to create online catalogue with high quality scans from their map collections. A good example is a global archive, the World Soil Survey Archive and Catalogue (WOSSAC) based in Cranfield University, UK, where soil maps are held in a series of map cabinets and map chests, filed by country and continent, and are still in the process of cataloguing [13]. Other significant collections exist worldwide, in Europe, USA and Russia. Each collection reflects the history and context of the institution which holds it. "A key aim of WOSSAC is to develop linkages with these other world databases to create a web of inter-connected land related information specifically soil surveys. There are tens of thousands of maps from all over the world. The process therefore of digitising and scanning the collection remains a major and significant task that will require significant funding" [12].

World Soil Information (ISRIC, Netherlands) has compiled a large collection with emphasis on the developing countries since 1966. "The ISRIC library has built up a collection of around 10000 (digitized) maps. The map collection contains mainly small-scale (1:250000 or smaller) maps. A significant part of the ISRIC map collection was scanned at the European Commission Joint Research Centre (JRC) as a foundation for the European Digital Archive of Soil Maps (EuDASM). The main objective was to transfer soil information into digital format, with the maximum resolution possible, to preserve the information of paper maps that are vulnerable to deterioration. The ISRIC library page has a powerful search function. A search can be done by author name, title words, keywords or country name, using basic and advanced search options. There is also the possibility to search on geographic coordinates. Some 80% of the maps are now digitized and these can be downloaded at high resolution and viewed on screen with a zoom facility. Scanning of ISRIC's library holdings is ongoing" [10,11,14].

Since the early 1950's FAO gave technical assistance on soils to member countries. "FAO Land and Water Division (NRL) has made an effort to make Soil Legacy data and information available for their users. In that regard, FAO has just finished uploading 1228 soil and land legacy maps (mainly soil maps and also land use, geological and land cover legacy maps). FAO will continue working in this activity and will include Soil Profile Legacy data soon. These maps were scanned as JPG and then uploaded with standard metadata (they were not geo-referenced)" [4].

In Switzerland the National Service Centre for Soil Mapping was disbanded in 1996. Today, soil information of sufficient quality and coverage has been collected for just 10 to 15% of agricultural land in NADOBAT soil information system [6]. "Yet this lack of standardised nationwide soil information is one fundamental reason why soil is scarcely considered in many policy areas and decision-making processes. National Research Programme started in 2018 "Sustainable Use of Soil as a Resource" (NRP 68) and is proposing a "Swiss soil information platform" showing how soil mapping can be expedited by using digital methods. Data collection and the creation of this soil information platform is considered a forward-looking investment that will span the next twenty years and will cost an estimated 15–25 million Swiss francs per year, primarily to pay for the necessary infrastructure and for mapping by private engineering offices. The authors use ten examples to illustrate how nationwide soil mapping creates added value. Thanks to improved soil information, the cost of drinking water treatment alone can be reduced by 10–15% or 7–10 million francs per year. This information will also help to reduce soil compaction and erosion, and optimise the use of fertilisers in farming. Very conservative assumptions for the ten examples indicate that the benefits of improved soil information across Switzerland will amount to between 55 and 132 million francs per year. "Every franc invested in soil mapping will therefore be repaid many times over, to the benefit of society and future generations" [7].

Other examples of ongoing work are British Library's Georeferencer interface which has 50 000 georeferenced historic maps placed on map of the world [2] with the visualization tool [3] and a variety of soil map galleries in the USA, such as United States Department of Agriculture [8], Minnesota map gallery [5], Historical Soil Survey Maps of South Carolina [9] and others.

However, none of these digital resources contain significant amount of soil maps for the territory of former USSR, and the software used and developed are not freely available.

3   Challenges

Manual processing of maps generates non-reproducible data (inaccuracy and subjectivity) and is not possible for handling large numbers of maps (thousands). The process of information extraction from map archives needs to be automated to the highest possible extent. Speaking about paper documents, optical character recognition (OCR) techniques are widely used for text digitization. However, probably because maps contain large amount of other symbols placed independently of the text, not a lot of studies show examples of OCR implementation on maps [1]. Metadata extraction from soil maps particularly is challenging due to difficult layouts and background noise. A soil map can contain within itself additional maps, text can be placed outside, as well as inside the map frame. Thus off-the-shelf OCR software is not suitable. We need to construct a pipeline of techniques for obtaining geo-referenced metadata from a high-resolution images of soil maps.

Without metadata extraction soil maps are browsable only by bibliographic information. However useful queries in actual research practice and, in particular, in prioritizing maps in its usability, and for scanning and detailed digitisation, is not bibliographic information, but metadata on actual contents.

"To understand further potential challenges in automating large-scale information extraction tasks for map archives, it is useful to efficiently assess spatio-temporal coverage, approximate map content at different map scales. Such preliminary analytical steps are often neglected or ignored in the map processing literature but represent critical phases that lay the foundation for any subsequent computational processes including recognition. Typically, knowledge about the variability in content and quality of map archives are a priori not available, since such large amounts of data cannot be analyzed manually. However, such information is critical for a better understanding of the data sources and the design of efficient and effective information extraction methods" [15].

On example of the soil map collection held in Dokuchaev Soil Science Institute archive, we demonstrate how a preliminary analyses can be systematically conducted to automate creation of a digital, visual, catalog of soil maps archive. The process is seen as semi-automatic requiring algorithmically-guided operator interactions.

Archive soil maps in this study are one of the sources of data in a Unifies Soil Information Analysis System (USIAS). USIAS data base includes a "digital map library" on the level of "data", and on the level of "procedures" it stores full protocols of map recognition procedures, similar to experimental data and analysis protocols. The scheme of USIAS formalization allows to store maps information with different level of detalization (starting from bibliographic and metadata to fully vectorized maps) in the same system. Formalized data sources and its producing protocols allow for automated data processing, accelerating the process of analysis, error detection, and additionally reduce uncertainty in model projections in USIAS by assigning weights (method precision, data source reliability, experimental or processing errors) to the data. To approach the two major goals of USIAS - reproducibility of soil research results and compatibility between standards—in the current study we present a tool for data collection from archive soil maps.

4   Obtaining Map Images

Maps archive consists of several thousands of sheets (in 285 shelves of the map depot) in different condition of paper, quality and completeness of output information. We used a test set of 207 maps (3 shelves) for this Demo.

Soil maps were photographed on a 2×1.5 m table with a mass-market camera of 24 Mpx resolution using a tripod with additional static diffused light. Images were obtained as jpeg files of about 10 Mb in size. For map sizes A5-A0 we recommend a camera with external finderscope for maximal use of camera resolution, lens with variable focal length of 24–70 mm (or 40–90 mm at 1.5 m installation height), frame supervision and wifi-interface for immediate image transfer to workstation. Image processing of the test set demonstrated high sensitivity to image quality. Experimental trials suggested the used resolution as a minimal required. For better performance of image processing algorithms it is recommended also to press down a map with a transparent (glass) cover to straighten corners and rolled maps.

5   Results

The presented map recognition tool is a part of USIAS database. This database allows us to store of heterogeneous data including field soil descriptions, images, maps, field and laboratory measurement results and archive data. Its advantageous specific features are the ability to store in addition to "data" information about how these data were obtained and processed, denoted as "objects preparation protocol", "data obtaining protocol" and "data processing protocol". This makes soil analysis reproducible and compatible with other standards.

1. "Object preparation protocol" is intended to formally describe map marking with a 3 symbol key (0–9 and A–Z symbols coding 42875 items) using a label printer.
2. "Data obtaining protocol" for maps is intended to formally describe the procedure of obtaining raw images by high-resolution photography (Sect. 4).
3. "Data processing protocol" for maps is intended to formally describe the pipeline of algorithms and its parameters (with scripts) for metadata extraction and georeferencing.

This pipeline of algorithms consisting of:

1. Image pre-processing steps:
   - removing a linear luminance trend;
   - correction of distortion due to perspective. It is done automatically using Python OpenCV: convert the resized image to grayscale, blur it and find edges. Next, find contours in the edged image, keep only the largest ones, from which choose the largest countour with 4 points, get corner coordinates and compute the perspective transform matrix and apply it convert the image. The operator inspects the result and if it is inappropriate,

chooses manual mode and points to the map corners in the image with a mouse and the perspective transform matrix is calculated and applied again;

- detecting a map frame by finding again the largest rightangled countour. Any text located outside map frame is highly probably an imprint;
- map language is detected using R package cld3. Further, whenever the detected language is not Russian (e.g. former USSR languages) a request is sent for translation of recognized text into Russian using Yandex translate API. In the test set 197 cases were detected as being in Russian or Ukrainian languages, 10 were unrecognized (due to low quality of text recognition in those cases);
- the area of a map itself and outside map area, which includes all titles, texts and legends are separated with a combination of image segmentation methods. First, we apply SLIC (Simple Linear Iterative Clustering) algorithm for Superpixel generation in color space and make a mask by adaptive thresholding and, next, using this mask that approximated the segmentation, we use GrabCut method (based on graph cuts) to accurately segment the foreground of an image, or the soil map itself, from the background Fig. 1; this preprocessing step is required for further separate image recognition of the map and outer parts. For example, location names can be on the map area as well as in the legend or surrounding texts, which will introduce an error into the automatic georeferencing. As well metadata extraction from texts and legends is performed better when it is not deteriorated by the colorful part of the map.

2. Rough soil map georeferencing is carried out using deep learning-based algorithms, first, Efficient and Accurate Scene Text Detector (EAST) for text detection within the map area, followed by text recognition using Tesseract OCR (LSTM mode) with the detected language to get locations names and their coordinates on the Fig. 2.

Then all obtained words are checked for being similar to a location name and geographical coordinates are obtained using Geonames database. Bounding box for the map is then calculated (minimum of 2 recognized locations is required) and the image is depicted on a global map (using leaflet) Fig. 3.

If the operator finds the result incorrect (obviously wrong map position) he is suggested to interactively point with a mouse to at least to locations and type in their names. Georeferencing is proceeded by requesting geographic coordinates for user input, calculating bounding box and placing the image on a globe. An example of user interface dialog is shown in Fig. 4.

3. Deriving soil map metadata is carried automatically with text recognition for the whole non-map area and provides possibility for keyword search. In addition, the operator can select map type (soil, parent material, vegetation and etc.), select condition of map paper and then select areas for text block recognition. Each area is processed using Tesseract OCR (LSTM mode) with the detected language and a translated (when necessary) text is shown to the operator together with selected part of the image. Operator is supposed to check the correctness of the text and choose/add attribute name for the

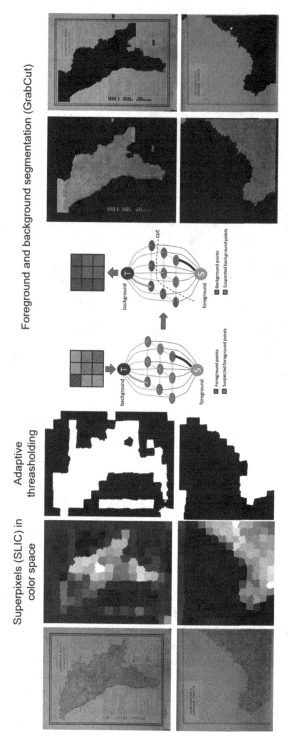

Fig. 1. Map segmentation into foreground (map) and background (with supplementary texts and legends) using a combination of methods in computer vision and neural networks

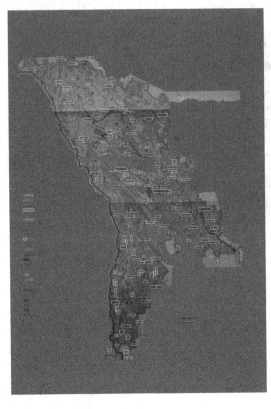

```
"n";"x";"y";"location_name"
"1";631;428;"СОКИРЯНЫ"
"2";1936;1863;"Слободу"
"3";1317;2742;"ВОЛГРАД"
"4";1906,5;2036,5;"Суворово"
"5";1961,5;2010,5;"Луржоры"
"6";1055,5;680,5;"СОРОНИ"
"7";1078,5;875;"ФЛОРЕШТЫ"
"8";1984,5;2057,5;"Олонешть"
"9";1529,5;942,5;"РЫБНИЦА"
"10";1619;1400;"ДУБОССАРЫ"
"11";1503;1014;"Воронково"
"12";1605,5;1493,5;"ошнида"
"13";1062,5;1092,5;"Ладувск"
"14";1734;1816;"БЕНДЕРЫ"
"15";854;981;"БЕЛЬЦЫ"
"16";1794,5;1924,5;"ЭККАУШАНЫ"
"17";1178,5;1200;"Теленешт"
"18";1905;1765,5;"ТИРАСПОЛЬ"
"19";1603,5;1974,5;"Тараклёя"
"20";942;944;"Бируимца"
"21";1723;1510,5;"риторяополь"
"22";1344;2562,5;"уанлия"
"23";1234,5;1521;"Стращены"
"24";1231;2686;"Говонойса"
"25";746;607;"аощёны"
"26";1412;1290;"ОРГЕЕВ"
"27";691,5;711;"гуФаны"
"28";1249;1104,5;"ззамещткы"
"29";1206;1288,5;""
"30";1124,5;1568;"оНмопорены"
"31";814;1140;"ФАЛЕШТЫ"
"32";1505;1607;"КИШИНЕВ"
```

Fig. 2. Example of obtaining a list of locations on a map image and corresponding coordinates. The text to the left of the picture is the resulting table containing: "n"—number of the location; "x", "y"—location coordinates on the image, "location name"—location name in Cyrillic.

block (title, scale, authors, organization, legend or etc.) and save metadata. Bounding boxes of all operator-selected and verified recognized text are saved and used as annotations for further use as a training set in layout detection techniques. An interface example is shown in Fig. 5.

Using this software demo 207 soil maps were added to USIAS. Suggested workstation and software demo provide sufficient speed of data processing to make the speed of catalog fill-up limited only by the time needed for photographing map sheets. While gained information will significantly accelerate the process of selective vectorization of maps in the next stage (scanning).

The user interface allows batch upload of images for georeferencing and annotation. Addition of soil maps into the USIAS is supposed in 3 stages: minimal (first stage), annotated (second stage) and detailed (third stage). First stage allows fast digital inventory of the library and rough georeferencing for the overview of quantity and geographical coverage, and a keywords search. The sec-

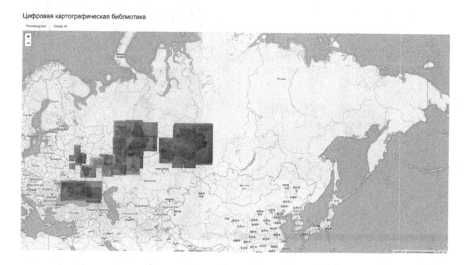

Fig. 3. Interactive tool for visualization of map images in high resolution on map of the world

Fig. 4. Interactive tool for rough georeferencing of map images in semi-auto mode by selecting two location points on a map with a mouse and entering location names. Geographic coordinates for the two selected and named points are returned by geonames.org and are displayed in gray boxes. Upon "save" these coordinates are used to recalculate the image bounding box and visualize it on the world map. (Color figure online)

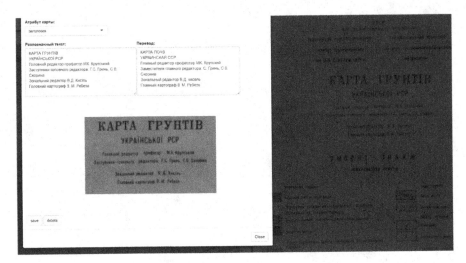

Fig. 5. Interactive tool for text recognition of metadata from selected map region with translation into Russian. In this example a map area containing title and authors was selected. In the upper line an attribute "header" was created which holds information about map title and authors. OCR result of original text is shown in the left box. Map language was Ukrainian, thus the result of its translation into Russian was given in the right box. This user-mode is proposed for the annotation stage, at which the user is supposed to select map areas, assign them attribute names and control/correct the results of OCR and translation in the interactive input box.

ond, annotated stage allows in addition parameter search (title, scale, authors, organization, type of soil and etc.), as well at some point it creates a training set for complex layout recognition for soil maps and will allow reduction of manual operations as stage two progresses. The quality of the results will and performance characteristics will be assessed at this stage. First two stages are supposed to produce a digital map catalog, summary overview and select maps for the 3rd stage. In the third stage we suppose to add scanned and vectorized maps into USIAS in the order of revealed priority. This stage is a continuous and longest process which is still carried out manually and requires automation to the highest degree possible using more advanced image recognition techniques. Currently developed demo tool can be adapted for OCR of map legends and explanatory notes and reports, template matching for estimation of areas, covered by various soil types and identification of a soil type in a requested point using the most large-scale available soil map.

All the software used in this research is available as Free and Open Source Software in public repositories.

6 Summary

To enable fast overview of a large Russian soil map collection and accelerate its further digitisation we propose a soil map recognition tool. This demo tool uses a pipeline of algorithms based on image processing with computer vision and neural networks applied to high resolution map photographs. This demo is a data collection tool developed for USIAS which aims for reproducibility in soil research and inter-connection of heterogeneous soil data.

References

1. A collection of abstracts of a special conference 2006–2019 "Digital Approaches to Cartographic Heritage". http://terkeptar.elte.hu/ch/?order=yd. Accessed 25 May 2021
2. British Library's Georeferencer tool. https://www.bl.uk/projects/georeferencer. Accessed 25 May 2021
3. British Library's Georeferencer tool. https://britishlibrary.georeferencer.com/api/v1/density. Accessed 25 May 2021
4. Food and Agriculture Organization of the United Nations (FAO), FAO soil and land legacy maps. http://www.fao.org/soils-portal/soil-survey/soil-maps-and-databases/fao-soil-legacy-maps/en/. Accessed 25 May 2021
5. Minnesota IT Services, Geospatial Information Office, Map Gallery, Geology and Soils. https://www.mngeo.state.mn.us/chouse/mapgallery.html. Accessed 25 May 2021
6. NADOBAT soil information system. https://www.nabodat.ch/index.php/de/service/bodenkartierungskatalog. Accessed 25 May 2021
7. Swiss National Science Foundation, Switzerland needs nationwide soil mapping. http://www.snf.ch/en/researchinFocus/newsroom/Pages/news-180419-press-release-nationwide-soil-mapping-for-switzerland.aspx. Accessed 25 May 2021
8. United States Department of Agriculture. https://www.nrcs.usda.gov/wps/portal/nrcs/detail/soils/survey/tools. Accessed 25 May 2021
9. University of South Carolina Libraries, DIGITAL COLLECTIONS. https://digital.library.sc.edu/collections/historical-soil-survey-maps-of-south-carolina/. Accessed 25 May 2021
10. World Soil Information (ISRIC), EuDASM - European Digital Archive of Soil Maps. https://www.isric.org/projects/eudasm-european-digital-archive-soil-maps. Accessed 25 May 2021
11. World Soil Information (ISRIC), Library and map collection. https://www.isric.org/explore/library. Accessed 25 May 2021
12. World Soil Survey Archive and Catalogue (WOSSAC), Archive Development. http://www.wossac.com/archive.cfm. Accessed 25 May 2021
13. Hallett, S.H., Bullock, P., Baillie, I.: Towards a world soil survey archive and catalogue. Soil Use Manag. **22**(2), 227–228 (2006). https://doi.org/10.1111/j.1475-2743.2006.00030.x, https://onlinelibrary.wiley.com/doi/abs/10.1111/j.1475-2743.2006.00030.x
14. Panagos, P., Jones, A., Bosco, C., Kumar, P.S.: European digital archive on soil maps (EuDASM): preserving important soil data for public free access. Int. J. Digit. Earth **4**(5), 434–443 (2011). https://doi.org/10.1080/17538947.2011.596580

15. Uhl, J., Leyk, S., Chiang, Y.Y., Duan, W., Knoblock, C.: Map archive mining: visual-analytical approaches to explore large historical map collections. ISPRS Int. J. Geo-Inf. **7**(4), 148 (2018). https://doi.org/10.3390/ijgi7040148, http://dx.doi.org/10.3390/ijgi7040148

16. Wadoux, A.M.J.C., Román-Dobarco, M., McBratney, A.B.: Perspectives on data-driven soil research. Eur. J. Soil Sci. 1–15 (2020). https://doi.org/10.1111/ejss.13071, https://onlinelibrary.wiley.com/doi/abs/10.1111/ejss.13071

Information Extraction from Text

Cross-Lingual Plagiarism Detection Method

Denis Zubarev[1,2]([✉]) [ID], Ilya Tikhomirov[4] [ID], and Ilya Sochenkov[1,2,3] [ID]

[1] Federal Research Center 'Computer Science and Control' of Russian Academy of Sciences, 44-2 Vavilov Str., Moscow 119333, Russia
zubarev@isa.ru
[2] Lomonosov Moscow State University, Moscow, Russia
[3] HSE University, Moscow, Russia
[4] Russian Science Foundation, Moscow, Russia

Abstract. In this paper, we describe a method for cross-lingual plagiarism detection for a distant language pair (Russian-English). All documents in a reference collection are split into fragments of fixed size. These fragments are indexed in a special inverted index, which maps words to a bit array. Each bit in the bit array shows whether a i_{th} sentence contains this word. This index is used for the retrieval of candidate fragments. We employ bit arrays stored in the index for assessing similarity of query and candidate sentences by lexis. Before doing retrieval, top keywords of a query document are mapped from one language to other with the help of cross-lingual word embeddings. We also train a language-agnostic sentence encoder that helps in comparing sentence pairs that have few or no lexis in common. The combined similarity score of sentence pairs is used by a text alignment algorithm, which tries to find blocks of contiguous and similar sentence pairs. We introduce a dataset for evaluation of this task - automatically translated Paraplag (monolingual dataset for plagiarism detection). The proposed method shows good performance on our dataset in terms of F1. We also evaluate the method on another publicly available dataset, on which our method outperforms previously reported results.

Keywords: Cross-lingual plagiarism detection · Cross-lingual word embeddings · Cross-lingual sentence embeddings

1 Introduction

Plagiarism is a serious and known problem in education and research, especially in developing countries like Russia. The availability of the huge amount of texts on the Web and free machine translation services makes it easier to create an "original" study. Systems for detecting plagiarism are very common now, and they are capable of detecting monolingual plagiarism even with obfuscations. However, detecting translated plagiarism is a challenging task. Commonly plagiarism detection is divided into two stages: source retrieval and text alignment.

© Springer Nature Switzerland AG 2022
A. Pozanenko et al. (Eds.): DAMDID/RCDL 2021, CCIS 1620, pp. 207–222, 2022.
https://doi.org/10.1007/978-3-031-12285-9_13

On the *source retrieval* stage for a given suspicious document, we need to find all sources of probable text reuse in a large collection of texts. For this task, a source is a whole text without details of what parts of this document were plagiarized. Typically we get a large set of documents (around 1000 or more) as a result of this stage. Those documents are called candidates.

On the *text alignment* stage we compare a suspicious document to each candidate to detect all reused fragments and identify their boundaries.

The same stages are valid for cross-language plagiarism detection. In this work, we describe a method for cross-lingual source retrieval and text alignment.

2   Related Work

The overview of different approaches for cross-lingual source retrieval is presented in [5] and [13]. Also, there made an evaluation and a detailed comparison of some featured methods. In [4], NMT (neural machine translation) is used to translate a query document to the other language. They solve the source retrieval task by employing shingles (overlapping word N-grams). To overcome machine translation ambiguity, they substitute each word by the label of the class. To obtain word classes, they apply agglomerative clustering on word embeddings learned from English Wikipedia. In a sentence comparison stage, they split sentences into phrases and employ phrase embeddings to find similar phrases in source documents. For training of phrase embeddings, they used an auto-encoder network for phrase restoration. Also, they generated similar phrases via double translation and employ margin-based loss to learn phrase similarity. In [8] is described a method for cross-language similarity detection based on the distributed representation of words (word embeddings). Experiments were conducted on English-French corpus. In [12] is described the training of word embeddings on comparable monolingual corpora and learning the optimal linear transformation of vectors from one language to another (there were used Russian and Ukrainian academic texts). Also, there were discussed usage of those embeddings in source retrieval and text alignment subtasks.

3   Cross-Lingual Plagiarism Detection Method

In this section, we describe a proposed method for Cross-lingual plagiarism detection. We selected methods for source retrieval and sentence comparison based on our previous works [23,24].

3.1   Preprocessing

On a preprocessing stage, we split each sentence into tokens, lemmatize tokens and parse texts. We train one Udpipe [19] model for both languages using a combination of Russian-syntagrus-ud-2.7 and English-ewt-ud-2.7 treebanks. Besides, we remove words with non-important part-of-speech tags: determiners, conjunction, particles (except not, no).

3.2 Source Retrieval

Firstly we split each document into fragments of maximum length L sentences with an overlap of O sentences. We index each fragment independently using a custom implementation of an inverted index. This index maps each word to a list of fragments that contain the word, and along with a fragment id, stores a binary array. i_{th} bit of a binary array is set to *true* if a i_{th} sentence of the corresponding fragment contains the word. Along with words, we index syntactic phrases up to 3 words.

We also split a query document into fragments and process each fragment in parallel. The top T terms are extracted from a fragment according to some weighting scheme. Then we map each keyword to N other language keywords with cross-lingual embeddings. If some keywords are already in another language, then they are added to the top without mapping. We preserve the weights of keywords from the original top. We describe the process of training cross-lingual embeddings in our previous work [23].

The searcher retrieves lists of fragments from the inverted index using query terms and merges them. Then we can use some ranking function to rank all fragments against query terms. We use well-known BM25 function: the score of a fragment F given a query Q which contains the terms $q_1, ..., q_n$ is

$$\text{score}(F, Q) = \sum_{i=1}^{n} \text{IDF}(q_i) \cdot \frac{f(q_i, F) \cdot (k_1 + 1)}{f(q_i, F) + k_1 \cdot (1 - b + b \cdot \frac{|F|}{avgdl})} \tag{1}$$

where $f(q_i, F)$ is q_i's term frequency in the fragment F, $|F|$ is the length of the fragment F in words, and $avgdl$ is the average fragment length in the collection. We use corresponding bit array F^{arr} to calculate an approximate term frequency of a term q_i in a fragment F: $f(q_i, F) = popcount(F_{q_i}^{arr})$, where *popcount* is a function that returns the number of bits set to *True* in a bit array. k_1 and b are free parameters (we use default values 1.2 and 0.75 appropriately). IDF is calculated by the following formula:

$$\text{IDF}(q_i) = \log_{10} \left(\frac{D_n - n(q_i) + 0.5}{n(q_i) + 0.5} + 1 \right) \tag{2}$$

where D_n is the total amount of documents in a collection, and $n(q_i)$ is the number of documents containing q_i.

We select the top K fragments with the highest score. We set $L = 64, O = 8, T = 100, N = 1, K = 300$ after tuning on dev dataset. In the next stage, we calculate similarity scores between sentences from the query fragment and selected fragments.

3.3 Sentence Similarity

We calculate similarity of a sentence pair as a weighted sum of two scores: a ratio of common lexis and sentence embeddings similarity.

$$sim(src, tgt) = W_l \cdot lexis\_score(src, tgt) + W_e \cdot cosine(emb(src), emb(tgt)) \quad (3)$$

We set W_l and W_e to 0.2 and 0.8, respectively, if they are not mentioned explicitly. In the next subsections, we describe how we calculate those sentence similarity scores.

Lexical Similarity. We compute the lexical similarity between two fragments right after candidate retrieval using bit arrays associated with each term. The algorithm of calculating lexical similarity is presented in the Listing 1. Each sentence pair is assigned a score, which is a sum of idf weights of terms that are shared by both sentences. This score is normalized by the sum of idf weights of all terms in the query sentence.

> **input** : *MaxSz* - max number of sentences per fragment, *query_terms* - a list
> of query term data structures, *candidate_terms* - a list of candidate
> term data structures
> **output:** *scores* - matrix of pairwise similarity between all sentences
> $scores = Array[MaxSz * MaxSz]$;
> $norms = Array[MaxSz]$;
> **for** $qterm \in query\_terms$ **do**
> > **for** $i \in 1..MaxSz$ **do**
> > > **if** $qterm.bitset[i]$ **then**
> > > > $norms[i] \mathrel{+}= qterm.idf$;
> > >
> > > **end**
> >
> > **end**
>
> **end**
> $matching\_terms = \{(q, c) \mid q \in query\_terms \land c \in candidate\_terms \land q.term\_id == c.term\_id\}$;
> **for** $(qterm, cterm) \in matching\_terms$ **do**
> > **for** $i \in 1..MaxSz$ **do**
> > > **for** $j \in 1..MaxSz$ **do**
> > > > $scores[i * MaxSz + j] \mathrel{+}=$
> > > > $qterm.bitset[i] * cterm.bitset[j] * qterm.idf / norms[i]$
> > >
> > > **end**
> >
> > **end**
>
> **end**

Algorithm 1. The algorithm of calculating lexical similarity between query and candidate sentences

query_terms can be the top T terms used in the retrieval stage. Also, merged posting lists used in the retrieval stage can be used as *candidate_terms*. In our experiments, we use all terms from the query fragment and all terms from each candidate. It requires additional read from a forward index to form *candidate_terms*, but K is usually small relative to the collection size, so it has little impact on performance.

Sentence Encoder Training. We train a sentence encoder for embedding sentences from both languages into a single semantic space.

Preparing Data. We use parallel sentences from various resources: Opus [20] (MultiUn, JW300, QED, Tatoeba), Yandex Parallel corpus [1], WikiMatrix [16], CCMatrix [17]. After downloading, all sentences are preprocessed, sentences of length < 4 words are dropped. We calculate hashes of sentences represented as concatenated normalized words. Then we delete all duplicate pairs, for which both hashes are equal. After that, we have around 83 million parallel sentences. We randomly sample 500k pairs as a development dataset.

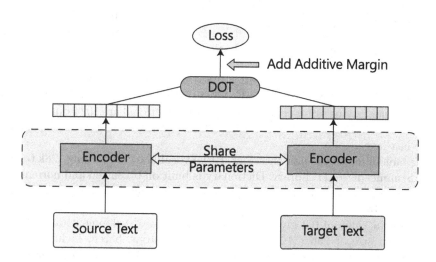

Fig. 1. Dual encoder architecture

Model. Following [7,22] we use a dual-encoder model consisting of paired encoders (see Fig. 1). The source and target sentences are encoded separately, but encoders within a language pair share parameters for both languages. We train the dual encoder on the task of translation retrieval: it is required to place tgt_i, the true translation of src_i, over all other target sentences. We add hard negatives for each src_n in every batch and also use all other sentences in the same batch except tgt_n as negative examples [9].

For source and target sentence embeddings x_i and y_i, the model can be optimized using the loss shown in formula 4.

$$L = -\frac{1}{N} \sum_{i=1}^{n} \frac{e^{\phi(x_i, y_i) - m}}{e^{\phi(x_i, y_i) - m} + \sum_{n=1, n \neq i}^{N} e^{\phi(x_i, y_n)}} \tag{4}$$

The final embeddings x_i, y_i are $l2$ normalized. We use dot-product as the function $\phi(x, y)$, which allows calculating cosine similarity for all sentences in a batch with single matrix multiplication. To improve training convergence, the cosine similarity score is scaled by the constant 20 (value was chosen empirically). Following [21,22] we also employ additive margin m, which increases the separation between translations and other sentences.

Hard Negatives Retrieval. We embed all target sentences using LASER [3] encoder and index them with Faiss library [10] using IvfPQ64 index (with 40k number of clusters trained on 40% of vectors). We search for the top 30 nearest neighbors (with nprobe = 64) of each source sentence. If a true translation sentence tgt_i is not found, then this pair of sentences is dropped (there were around 2.5m such sentences). After that for each src_i we select 4 sentences that satisfy the condition: $\{t \mid t \in top\_sim(src_i) \wedge sim(t) < sim(tgt_i) - 0.035 \wedge hns[t] < 500\}$ where hns is a global counter of all hard negatives, introduced to limit the number of times a sentence can be a hard negative.

Encoder Architecture and Configuration. We use stacked BiLSTM with three layers - each 512-dimensional, input embedding size is set to 320. The resulting sentence representations (after concatenating both directions) are 1024-dimensional. Sentence embeddings are obtained using a max-pooling strategy over the output of the last layer.

We build a joint byte-pair encoding (BPE) vocabulary containing 128k tokens using Sentencepiece [11] library. Dictionary is built on tokenized and normalized sentences of both languages, so we enable only splitting by whitespaces in Sentencepiece library.

A margin value of 0.2 gives the best results on a development dataset. We use a batch size of 256 (˜1280 sentences in the target batch, including hard negatives). The learning rate is linearly warmed up to 0.001 (during 250k steps) and then reduced using an inverse square root schedule. We stop training at 2.5 m steps (8 epochs). We also follow [6] and multiply the gradients of the BPE embeddings by a factor of 15. It slightly improves accuracy on the test data according to our experiments.

Sentence Embeddings Similarity. We precompute sentence embeddings for all sentences in reference collection and compress them with PQ64 encoding so they can fit into memory. At the search time, query sentences are encoded on the fly while sentences from K selected fragments are fetched from memory. Then we use brute-force nearest neighbor search to find k most similar sentences for each query.

3.4   Text Alignment Algorithm

In this stage, scored sentence pairs found for all query fragments are merged and grouped by a candidate document. We work with the pair of query and candidate documents in this stage. The main goal of a text alignment algorithm is to find blocks of contiguous and similar sentence pairs. Moreover, it is necessary to unambiguously link each sentence of the query document with one or more sentences from the candidate document or mark it as the original. According to [15] typical text alignment algorithm consists of four stages: preprocessing, seeding, extension, and filtering. We will describe the three last stages in this section.

We use sentence pairs with the similarity score larger than sim_{seed} as seeds. Given the seeds, the task of the extension stage is to form larger text blocks that are similar between two documents. We will define block b as the set of sentence pairs SP. We introduce some functions to simplify the description of the extension algorithm: $querySents(b)$ - returns all query sentences № included in the given block, $firstSentQ(b), lastSentQ(b)$ - returns № of the first and last query sentences of the given block, $firstSentCand(b), lastSentCand(b)$ - returns № of the first and last candidate sentences of the given block. In the beginning, each seed s_i is assigned to its block b_i. Then we iterate over all query sentences that are placed after and before the seed in the query document. We create a new block $\hat{b}$ by adding a sentence pair sp to the block b if

$$firstSentQ(b) - qgap \leq sp.query\_num \leq lastSentQ(b) + qgap \qquad (5)$$

and

$$firstSentQ(b) - qgap \leq sp.query\_num \leq lastSentQ(b) + qgap \qquad (6)$$

and similarity of sp is greater than sim_{min}. We only keep blocks that cover more than $block\_min\_size$ query sentences or contain a sentence pair with a similarity score greater than sim_{high}. After the extension, step there is a set of blocks B, and there may be blocks that share the same query sentences. It is required to find a subset of B called partition p_{goal} that holds this property:

$$p_{goal} = \{b \mid b \in B \wedge \forall i,j : querySents(b_i) \cap querySents(b_j) = \varnothing\} \qquad (7)$$

and covers as many query sentences as possible. We search over all possible partitions using a variation of beam search. We start with an empty partition and create new partitions by iterating over B and adding a block b to the current partition if it does not violate the property. After that, we select only MP most promising partitions and repeat until we get partitions, which can not be extended by any $b \in B$. We rank partitions using the function shown in 8.

$$score(p) = 0.6 \cdot \log_T \left(\left| \bigcup_{sp \in p.pairs} sp.q \right| \right) + 0.4 \cdot \log_T \left(\sum_{sp \in p.pairs} sim(sp) \right) \qquad (8)$$

where T is a total count of query sentences, $sp.q$ is sentence № in a query document, $sp.pairs$ is a set of all sentence pairs from all blocks included in a partition p. We set $qgap = 2, cgap = 3, sim_{seed} = 0.45, sim_{min} = 0.28, block\_min\_size = 2, sim_{high} = 0.5, MP = 5$ after tuning on dev dataset. Also, we remove documents, which contain less than three reused sentences.

4   Evaluation

4.1   Essays Dataset

We translated the text of sources from the Paraplag dataset from Russian to the English language[1]. Paraplag [18] is the monolingual dataset for evaluation of plagiarism detection methods. This dataset contains manually written essays on the given topic. Authors of essays should have used at least five sources, which they had to search by themselves when composing essays. The plagiarism cases vary by the level of complexity: from copy-paste plagiarism to heavily disguised plagiarism. We evaluate the BLEU score of various translation services for Russian → English translation on a test dataset (Table 1). We used 300 sentences from WMT-News dataset[2], 300 sentences from News-Commentary dataset[3] and 300 sentences from cross-lingual essays.

Table 1. BLEU scores of popular MT services.

Translation services	BLEU
Google translate	37.9
Yandex translate	38.61
Yandex Cloud translate	40.32

In the end, we used Yandex Cloud translate API to translate 1400 sources. It should be noted that we translated only sentences that were reused in essays and their context (9500 characters above and below). The statistic of the translated dataset is presented in the Table 2.

Table 2. Statistic of essays dataset

	Essays			Sources		
	# of essays	Avg # of chars	Avg # of sents	# of sources	Avg # of chars	Avg # of sents
Essay1	114	17741	162	742	14318	112
Essay2	105	13371	110	682	15277	120

Essay1 is a dataset that consists of copy-pasted/moderately disguised essays, whereas essay2 contains only heavily disguised essays. We randomly sampled 30 essays of each type as a development dataset and used the rest essays as test data.

---

[1] https://plagevalrus.github.io/content/corpora/paraplag_v2.html.

[2] http://opus.nlpl.eu/WMT-News.php.

[3] http://opus.nlpl.eu/News-Commentary.php.

Examples of some plagiarism cases from both datasets are presented in Tables 3 and 4.

Table 3. Samples of reused text from Essay1 dataset

Reused sentence	Sentence from a source
После опубликования романа «Атлант расправил плечи» Рэнд посвятила остаток жизни разъяснению своих философских взглядов.	After the publication of the novel "Atlas Shrugged", Rand devoted the rest of her life to explaining her philosophical views.
У Генри Джеймса были основания видеть в молодом Киплинге будущего английского Бальзака, потому как он открыл героику повседневных будней н романтизировал обыденность.	He discovered the heroics of everyday life and romanticized the everyday.
Воды породили огонь и с помощью великой силой тепла в них рождено было Золотое Яйцо.	The Golden Egg was born in them by the great power of heat.
Возникает вопрос, почему свойства воды могут изменяться и восстанавливаться при переходе в разные агрегатные состояния?	Why can the properties of water change and recover during the transition to different aggregate states?
Судьба этого человека неизвестна, но дело его остается открытым на протяжении вот уже 40 лет.	The fate of this man is unknown, but his case has remained open for 40 years, and during this time 60 volumes of assumptions and versions have been accumulated.

4.2   Evaluation Results

We use standard for this tasks metrics: micro-averaged precision and recall at a character level, granularity, plagdet. Granularity penalizes detectors for reporting overlapping or multiple detections for a single plagiarism case. Plagdet combines F1-score and granularity into a single overall score.

We indexed all articles from English Wikipedia dumps (March 2020, more than 6M articles) and essay sources into one collection using $L = 64$ and $O = 8$. When splitting query documents, we also used fragments of size 64 (L) but without overlap ($O = 0$). There were 10 parallel threads while searching in an inverted index and 2 parallel threads when calculating a similarity between sentences. Tests were run on common hardware: Core i9-9900K, 64 GiB RAM, 2 GPU GeForce RTX 2080. Table 5 displays the evaluation results obtained on both datasets.

We used the same set of parameters for both datasets. These results show that the method works well for copy-pasted/moderately disguised translations, and it struggles to detect heavily disguised reused translated text. Since the test data (Essay1) was annotated with the type of obfuscation used when modifying

Table 4. Samples of reused text from Essay2 dataset

Reused sentence	Sentence from a source
Платон послужил источником некоторых идей для Зигмунда.	Being familiar with Plato's philosophical ideas, Freud undoubtedly drew from there some ideas about the unconscious.
Ее партнером по фильму стал Джонни Ли Миллер, который в будущем станем ее первым мужем.	And in 1995, she starred in her first major project, the film "Hackers", which was partnered with Johnny Lee Miller, who later became her husband.
Они благодарны за все блага, которые у них есть, вне зависимости от их величин мало или много не важно.	Successful people are grateful for all the benefits that they have, regardless of their size.
Фильм «Космическая одиссея 2001» принёс Стэнли Кубрику мировую славу, а сенсацию вызвал «Заводной апельсин», который запретили к показу в Великобритании.	The film "A Clockwork Orange" (1971), based on the novel of the same name by Anthony Burgess, became truly sensational.
Мерфи больше думал о играх сборной Уэльса, нежели о матчах Кубка европейских чемпионов в Югославии, так как команда проходила в полуфинал и это занимало все его мысли.	To be honest, I was thinking much more about our matches at that moment than about the match in Yugoslavia.

Table 5. Evaluation results.

Dataset	Rec	Prec	Gran	Plagdet	Time (s)
Essay1	0.907	0.839	1.06	0.838	29.24
Essay2	0.6	0.754	1.05	0.645	22.31

Table 6. Recall for each obfuscation type

Type	Description	Recall
CCT	Concatenation of sentences	0.851
HPR	Heavy modifications (paraphrasing)	0.811
SSP	Splitting of sentences	0.871
LPR	Minor modifications (replacing/reordering of words etc.)	0.942
ADD	Addition of words	0.936
DEL	Deletion of words	0.93
CPY	Copy/paste	0.945

each sentence, we were able to identify the most difficult types of obfuscation for the method. Recall for each obfusctation type is shown in the Table 6.

This result shows that the most challenging type of obfuscation for our method is heavy paraphrasing and concatenation/splitting of sentences. Concatenation/splitting of sentences is hard to detect with our method when sentence parts have uneven sizes.

4.3 Comparison with Other Methods

We also compare our method with a method proposed in [4], which was described in related work. Their dataset was automatically generated. The reference English collection contains about 100k randomly chosen Wikipedia articles. Query documents were sampled from Russian Wikipedia, each document was translated into English, and 500 most similar documents were retrieved from the reference collection using $tf \cdot idf$. Then 20 to 80% of sentences in a query document were replaced by the machine-translated sentences from some similar documents. The test and dev datasets contain 316 and 100 documents, respectively. The metrics obtained on the test dataset and results from the paper [4] are presented in the Table 7.

Table 7. Comparison with other method

Method	Rec	Prec	F1	Gran	Plagdet	Time (s)
[4]	0.79	**0.83**	0.8	–	–	–
Default (64/8)	0.89	0.76	0.82	1.09	0.771	206.34
Tuned (64/8)	0.887	0.822	0.853	**1.08**	0.808	206.88
Tuned (16/4)	**0.924**	0.824	**0.871**	1.08	**0.825**	385.87

Default is a configuration tuned on essay datasets with the size of fragments $L = 64$ and the overlap $O = 8$. We slightly tuned default configuration to increase precision on the dev dataset (setting $sim_{min} = 0.32$, $sim_{high} = 0.6$, $sim_{seed} = 0.55$). Also, more fine-grained splitting of query documents into fragments of size 16 gives better recall. We don't observe this behavior on the essays dataset, presumably because essays are dedicated to one topic. In this dataset, random sentences may be injected into a query document from slightly related sources.

4.4 Fragment Size Experiments

We conduct experiments using various configurations of fragment sizes used in indexing and searching.

Retrieval Quality. Firstly we evaluate the impact of indexing fragments of smaller size on the task of fragment retrieval. This task is formulated as follows: split a given document into fragments and find all gold fragments. We use described Essays dataset for this task and employ source documents split into fragments as gold data. Micro/macro-averaged recall is used as a metric for this task. Results for some configurations are shown in the Table 8. The format of 'Query' and 'Idx' fields is L/O, where L is fragment size in sentences and O is an overlap with the previous fragment. T is the top size of terms, and K is the maximum amount of fragments to retrieve.

Table 8. Impact of the fragment size on the retrieval quality

Query	Idx	T	K	Micro	Macro	Time (s)
16/4	64/8	50	200	**0.889**	**0.912**	21.58
16/0	64/8	50	200	0.881	0.904	16.68
64/8	64/8	100	300	0.862	0.891	12.40
64/0	64/8	100	300	0.860	0.889	11.41
16/4	64/0	50	200	**0.871**	**0.891**	21.52
16/0	64/0	50	200	0.863	0.883	16.64
64/8	64/0	100	300	0.845	0.869	12.59
64/0	64/0	100	300	0.841	0.867	11.41
16/4	32/4	50	200	**0.831**	**0.857**	22.28
32/4	32/4	50	200	0.796	0.831	14.94
32/0	32/4	50	200	0.785	0.822	13.37
16/4	32/0	50	200	**0.827**	**0.851**	21.74
32/4	32/0	50	200	0.793	0.825	14.86
32/0	32/0	50	200	0.785	0.817	13.52

It is clear that recall steadily decreases when making indexed fragments smaller. At the same time, an overlap of indexed fragments can slightly improve retrieval performance. Reducing the size of the fragments of a query document has a positive effect on the quality of the retrieval, although it requires more computation time.

Plagiarism Search Quality. In this experiment, we fix the indexed fragments configuration: 64/8 and $T = 100, K = 300$. We use described Essays dataset and run the plagiarism detection method using various splitting configurations. Results of this experiment are shown in the Table 9. Queries configurations with star have different parameters $T = 50, K = 200$.

Table 9. Impact of the query fragment size on the search quality

Query	Dataset	Rec	Prec	F1	Time (s)
64/8	Essays1	0.811	0.722	0.764	27.15
64/0	Essays1	0.81	0.726	0.766	25.15
32/4	Essays1	0.809	0.729	**0.767**	35.83
32/0	Essays1	0.803	**0.734**	**0.767**	32.21
16/4*	Essays1	**0.814**	0.713	0.76	46.87
16/0*	Essays1	0.811	0.728	**0.767**	37.05
64/8	Essays2	0.501	0.588	0.541	29.46
64/0	Essays2	0.5	0.588	0.54	28.02
32/4	Essays2	0.498	0.593	0.541	40.62
32/0	Essays2	0.496	**0.596**	0.541	35.7
16/4*	Essays2	**0.514**	0.586	**0.548**	52.35
16/0*	Essays2	0.508	0.595	**0.548**	40.77

The impact of query fragment size is not as critical as it was for the retrieval task. Reducing the size of query fragments only marginally increases F1-score, whereas computation time is almost doubled.

4.5 Impact of Text Alignment Algorithm

In this experiment, we compare the quality of plagiarism detection with and without the use of the text alignment algorithm. We turn off text alignment algorithm by setting $qgap = 0$, $cgap = 0$, $block\_min\_size = 1$, then every sentence forms a single block. We also disable resolving conflicting blocks in this mode. We compare influence of text alignment using various sentence similarity scoring schemas:

- only-lexical - measure only lexical similarity between sentences. It is equivalent to these parameters: $W_l = 1.0$ and $W_e = 0.0$.
- only-emb - measure only similarity between sentence embeddings. It is equivalent to these parameters: $w_l = 0.0$ and $w_e = 1.0$.
- default - use linear combination of two similarity scores. It is equivalent to these parameters: $w_l = 0.2$ and $w_e = 0.8$.

We don't optimize parameters for each combination and use default parameters specified in this paper. Results of this experiment are shown in the Table 10.

Table 10. Impact of text alignment on the plagiarism detection quality

TA	Sim score	Dataset	Rec	Prec	Gran	Plagdet
On	Only-lexical	Essay1	0.879	0.86	1.06	0.835
On	Only-emb	Essay1	0.902	0.785	1.06	0.808
On	Default	Essay1	**0.907**	0.839	1.06	**0.838**
Off	Only-lexical	Essay1	0.46	0.853	**1.04**	0.58
Off	Only-emb	Essay1	0.809	0.824	**1.04**	0.795
Off	Default	Essay1	0.789	**0.882**	**1.04**	0.811
On	Only-lexical	Essay2	0.518	**0.777**	1.06	0.598
On	Only-emb	Essay2	**0.617**	0.711	1.05	0.639
On	Default	Essay2	0.6	0.754	1.05	**0.645**
Off	Only-lexical	Essay2	0.176	0.714	1.03	0.276
Off	Only-emb	Essay2	0.421	0.715	**1.02**	0.522
Off	Default	Essay2	0.364	0.775	**1.02**	0.488

Text alignment algorithm significantly boosts the performance of plagiarism detection. It is most clearly manifested for the only-lexical scoring scheme where recall is almost doubled for Essay1 and tripled for Essay2. It can be explained considering the structure of the reused text in the test data. Most reused sentences from the one source form contiguous text block that is easily identified by the text alignment algorithm. It just requires that there were one or more sentences with high similarity somewhere in this text block. It is questionable whether this reused text structure exhibits the real-world scenario. The essays were written by students, who were "allowed" to reuse text from multiple sources, and they chose this easy structure to compose their works. Thus it can be assumed that this structure is used in certain circumstances, and we even don't need sentence embeddings to detect reused text with light forms of obfuscation. Employing sentence embeddings similarity score works quite well by itself without text alignment algorithm, but enhancing with this algorithm gives higher recall.

5   Conclusion

In this article, we described a method for cross-lingual plagiarism detection for Russian-English language pair. We translated Paraplag to evaluate the proposed method on essays composed of reused text with varying levels of obfuscation. Conducted experiments showed that the method is viable and it achieves good performance in terms of F1 and plagdet on this dataset. In future work, we will concentrate on training sentence encoders and word embeddings on academic papers. There are some challenges to be solved, especially regarding the lack of parallel sentences of this thematic. One of the possible solutions is to mine parallel sentences using a trained sentence encoder from comparable corpora [16,

25]. Another possibility is to employ the generalization ability of state-of-the-art unsupervised multilingual models [2] and fine-tune them on scarce available data [14].

Acknowledgement. The reported study was funded by RFBR according to the research projects no. 18-29-03187 & no. 18-29-03086. The research is also partially supported by the Ministry of Science and Higher Education of the Russian Federation according to the agreement between the Lomonosov Moscow State University and the Foundation of project support of the National Technology Initiative No 7/1251/2019 dated 15.08.2019 within the Research Program "Center of Big Data Storage and Analysis" of the National Technology Initiative Competence Center (project "Text mining tools for big data").

References

1. Antonova, A., Misyurev, A.: Building a web-based parallel corpus and filtering out machine-translated text. In: Proceedings of the 4th Workshop on Building and Using Comparable Corpora: Comparable Corpora and the Web, pp. 136–144 (2011)
2. Artetxe, M., Ruder, S., Yogatama, D.: On the cross-lingual transferability of monolingual representations. arXiv preprint arXiv:1910.11856 (2019)
3. Artetxe, M., Schwenk, H.: Massively multilingual sentence embeddings for zero-shot cross-lingual transfer and beyond. Trans. Assoc. Comput. Linguist. **7**, 597–610 (2019)
4. Bakhteev, O., Ogaltsov, A., Khazov, A., Safin, K., Kuznetsova, R.: CrossLang: the system of cross-lingual plagiarism detection. In: Workshop on Document Intelligence at NeurIPS 2019 (2019)
5. Barrón-Cedeño, A., Gupta, P., Rosso, P.: Methods for cross-language plagiarism detection. Knowl.-Based Syst. **50**, 211–217 (2013)
6. Chidambaram, M., et al.: Learning cross-lingual sentence representations via a multi-task dual-encoder model. arXiv preprint arXiv:1810.12836 (2018)
7. Feng, F., Yang, Y., Cer, D., Arivazhagan, N., Wang, W.: Language-agnostic BERT sentence embedding. arXiv preprint arXiv:2007.01852 (2020)
8. Ferrero, J., Agnes, F., Besacier, L., Schwab, D.: Usingword embedding for cross-language plagiarism detection. arXiv preprint arXiv:1702.03082 (2017)
9. Guo, M., et al.: Effective parallel corpus mining using bilingual sentence embeddings. arXiv preprint arXiv:1807.11906 (2018)
10. Johnson, J., Douze, M., Jégou, H.: Billion-scale similarity search with GPUs. IEEE Trans. Big Data **7**, 535–547 (2019)
11. Kudo, T., Richardson, J.: SentencePiece: a simple and language independent subword tokenizer and detokenizer for neural text processing. arXiv preprint arXiv:1808.06226 (2018)
12. Kutuzov, A., Kopotev, M., Sviridenko, T., Ivanova, L.: Clustering comparable corpora of Russian and Ukrainian academic texts: word embeddings and semantic fingerprints. arXiv preprint arXiv:1604.05372 (2016)
13. Potthast, M., Barrón-Cedeño, A., Stein, B., Rosso, P.: Cross-language plagiarism detection. Lang. Resour. Eval. **45**(1), 45–62 (2011)

14. Reimers, N., Gurevych, I.: Making monolingual sentence embeddings multilingual using knowledge distillation. In: Proceedings of the 2020 Conference on Empirical Methods in Natural Language Processing. Association for Computational Linguistics (2020). https://arxiv.org/abs/2004.09813

15. Sanchez-Perez, M.A., Sidorov, G., Gelbukh, A.F.: A winning approach to text alignment for text reuse detection at PAN 2014. In: CLEF (Working Notes), pp. 1004–1011 (2014)

16. Schwenk, H., Chaudhary, V., Sun, S., Gong, H., Guzmán, F.: WikiMatrix: mining 135M parallel sentences in 1620 language pairs from Wikipedia. arXiv preprint arXiv:1907.05791 (2019)

17. Schwenk, H., Wenzek, G., Edunov, S., Grave, E., Joulin, A.: CCMatrix: mining billions of high-quality parallel sentences on the web. arXiv preprint arXiv:1911.04944 (2019)

18. Sochenkov, I., Zubarev, D., Smirnov, I.: The ParaPlag: Russian dataset for paraphrased plagiarism detection. In: Computational Linguistics and Intellectual Technologies: Papers from the Annual International Conference "Dialogue", vol. 1, pp. 284–297 (2017)

19. Straka, M., Hajic, J., Straková, J.: UDPipe: trainable pipeline for processing CoNLL-U files performing tokenization, morphological analysis, pos tagging and parsing. In: Proceedings of the Tenth International Conference on Language Resources and Evaluation (LREC 2016), pp. 4290–4297 (2016)

20. Tiedemann, J.: Parallel data, tools and interfaces in OPUS. In: Lrec, vol. 2012, pp. 2214–2218 (2012)

21. Wang, F., Cheng, J., Liu, W., Liu, H.: Additive margin softmax for face verification. IEEE Signal Process. Lett. **25**(7), 926–930 (2018)

22. Yang, Y., et al.: Improving multilingual sentence embedding using bi-directional dual encoder with additive margin softmax. arXiv preprint arXiv:1902.08564 (2019)

23. Zubarev, D., Sochenkov, I.: Comparison of cross-lingual similar documents retrieval methods. In: Sychev, A., Makhortov, S., Thalheim, B. (eds.) DAMDID/RCDL 2020. CCIS, vol. 1427, pp. 216–229. Springer, Cham (2021). https://doi.org/10.1007/978-3-030-81200-3_16

24. Zubarev, D., Sochenkov, I.: Cross-language text alignment for plagiarism detection based on contextual and context-free models. In: Proceedings of the Annual International Conference "Dialogue, vol. 1, pp. 799–810 (2019)

25. Zweigenbaum, P., Sharoff, S., Rapp, R.: Overview of the third BUCC shared task: spotting parallel sentences in comparable corpora. In: Proceedings of 11th Workshop on Building and Using Comparable Corpora, pp. 39–42 (2018)

Methods for Automatic Argumentation Structure Prediction

Ilya Dimov[1(✉)] and Boris Dobrov[1,2(✉)]

[1] Lomonosov Moscow State University, Moscow, Russia
iliyadimov@icloud.com, dobrov_bv@mail.ru
[2] Research Computing Center of M.V. Lomonosov Moscow State University,
Moscow, Russia

Abstract. Argumentation mining is a natural language understanding task consisting of several subtasks: relevance detection, stance classification, argument quality assessment and fact checking. In this work we propose several architectures for the analysis of argumentative texts based on BERT. We also show, that models, which jointly learn argumentation mining subtasks outperform pipelines of models trained on a single tasks. Additionally we explore transfer learning approach based on pretraining for the natural language inference task, which achieves highest score on tasks of argumentation mining among the models trained on english corpora.

Keywords: Argumentation mining · Relevance detection · Stance classification

1 Introduction

Argumentation is the process of forming reasons and of drawing conclusion and applying them to a case in discussion. The task of argumentation mining is aimed at the extraction of the argument components and linking them in a structure.

There are various ways to construct the argument structure, varying in the number of components and types of connections between them. In this work, following the notion introduced in the works of IBM Project Debater[1] [8,11,19], we use three entities:

- *Topic* - a short, concise and controversial statement, which serves to frame the discussion.
- *Claim* - a statement, that directly supports or contests the topic.
- *Evidence* - a statement, that directly supports or contests the topic, which is not a mere belief or a claim.

We additionaly define a *focus* as the subject mentioned in the *topic*. There are two tasks, that help to restore the structure of the argument:

---

[1] https://www.research.ibm.com/haifa/dept/vst/debating_data.shtml.

© Springer Nature Switzerland AG 2022
A. Pozanenko et al. (Eds.): DAMDID/RCDL 2021, CCIS 1620, pp. 223–233, 2022.
https://doi.org/10.1007/978-3-031-12285-9_14

1. Relevance detection - given a topic and a sentence, find out whether the sentence can be used as claim or evidence.
2. Stance classification - given a topic and a valid claim or evidence, classify the stance as a support or rebuttal (PRO/CON).

These tasks are hard to solve, as the texts often do not provide full context of the underlying discussion. For example, let's take the following pair of a topic and a claim:

```
Topic: We should further exploit nuclear power
Claim: Similarly, after the Chernobyl incident, ... they
believed a similar meltdown was likely to happen in the U.S.
```

There is no lexical intersection between the topic and the claim. To correctly link those two entities the system must understand, that Chernobyl is connected to nuclear energy. One way to solve this problem is to supply the required facts to the model as an additional input. Another approach is to use a large pretrained model. BERT [7] has shown great success in various tasks of natural language understanding such as question answering and commonsense reasoning [6].

In this work we explore several architectures based on BERT and a transfer learning approach and try to analyze the model outputs and understand, how to further improve its results. The code for our experiments is publically available[2].

2   Related Work

The task of argument mining was explored by a number of various approaches. For example systems MARGOT [12] and Targer [5] the task is formulated as a sequence tagging problem. MARGOT's pipeline first classifies sentences as containing claim or evidence with respect to a topic, before extracting a continuous span of text, which serves as claim or evidence. Similiar task is solved in [2], where a collection of phrases classified as argumentative or non-argumentative. Sequence-tagging approach is harder than sentence-level classification, because it requires a more complex dataset, which is difficult to construct.

The paper [15] proposes another corpora, where sentences are classified with respect to the focus as supporting, opposing and not relevant arguments.

In the work [1] the structure of the argument is different: evidence is connected to claims and claims are connected to topics, instead of the claim and the evidence being both connected to the topic. The proposed dataset has additional markup of evidence to 3 types, based on its origin: Anectodal, Expert and Study.

In [11,14] authors describe a method of automatic claim mining. They gather a weak labeled claim corpora and show, that it can be used for model improvement alongside human-annotated dataset.

Another subtask of argumentation mining is argumentation quality assessment [9,10,18]. The task is, given a claims or evidence relevant to the topic, to arrange them in the way that the most strong arguments are first. The are two

---

[2] https://github.com/hawkeoni/Argument-structure-prediction.

main approaches: first focuses on setting a value from 0 to 1 for an argumentation component, while the second approach consists of selecting the strongest claim or evidence phrases with respect to the topic out of several options.

As stated in the definition, evidence must not be a mere belief or a baseless statement. For example study-type evidence may often refer to some research, but such research may not exist, thus fooling the system. To battle this the task of fact checking is introduced. In CLEF-2020 CheckThat! [3] lab several subtasks are proposed, based on the data from Twitter:

- Check Worthiness - given a topic and a stream of potentially-related tweets, rank the tweets according to their check-worthiness for the topic. A check-worthy tweet is a tweet that includes a claim that is of interest to a large audience (specially journalists), might have a harmful effect, etc.
- Claim Retrieval - Given a check-worthy claim and a dataset of verified claims, rank the verified claims, so that those that verify the input claim (or a sub-claim in it) are ranked on top.
- Evidence Retrieval - Given a check-worthy claim on a specific topic and a set of text snippets extracted from potentially-relevant webpages, return a ranked list of evidence snippets for the claim. Evidence snippets are those snippets that are useful in verifying the given claim.

Similar problem is stated in FEVER dataset [17]: given a claim the task is to find relevant Wikipedia sentences, which support it.

Another similar task is natural language inference or textual entailment. Given a text and a hypothesis the goal is to determine, whether the hypothesis is true. While the task is not directly connected to argumentation mining it is highly similar to it. Hypothesis can be entailed by text, contradict it or not be relevant to the text, just as claims or evidence can be relevant and not relevant to the topic.

3   Model Architectures

We formulate tasks of stance classification and relevance detection as binary classification problems and solve them with three architectures, based on the english version of BERT-base model from [7]. All architectures take a pair sentences as input: a topic and a claim or a topic and evidence, and then predict the probability of those components being related, or the probability of the support relation depending on the task.

BERT is a transformer encoder [20] built on the self-attention mechanism. We use $BERT_{base}$ model weights, which consist of 12 transformer blocks. Each block calculates multihead self attention over the inputs (either token embeddings or previous layer outputs):

$$Attention(Q, K, V) = softmax(\frac{QK^T}{\sqrt{d_k}})V \qquad (1)$$

where Q, K, V are linear projections of the inputs, d_k is the dimension of attention head. The self-attention layer is followed by a layer normalization of sum of the input and the resulting self attention activations, which is called residual connection. Normalized activations are fed to a feedforward layer, which adds nonlinearity to the model:

$$FeedForward(x) = max(0, xW_1 + b1)W_2 + b2 \qquad (2)$$

The feedforward layer is followed by another normalization over a residual connection.

The original BERT model is trained on a masked language model task, where it has to predict masked tokens in the sentence by leveraging the context from all tokens, as opposed to RNN-based models, which employ either the left or the right context or rely on bidirectional approach where to RNNs are used. Additionaly authors use the task of next sentence prediction, where the model also predicts, if two sentences follow each other.

The task of masked language modeling provides a good starting point for model finetuning of various downstream tasks such as named entity recognition, natural language inference, question answering and many other, as stated in the original work [7].

We formulate the subtasks of argumentation mining as classification and solve them the following way: we encode pairs of sentences (topic and claim, topic and evidence) via BERT and use this rich contextual representation to classify the relation between texts.

Our first architecture $BERT_{basic}$ follows a standart approach for classification originally used for next sentence prediction. We concatenate the text in the following way:

$$[CLS]\ text_1\ [SEP]\ text_2\ [SEP] \qquad (3)$$

where [CLS] is a special token representing the start of the text and [SEP] represents the end of the text. The final vector representation of the [CLS] token is used in a linear layer followed by a softmax over classes for classification. This is a basic approach for text classification with BERT, which makes the model learn a suitable representation of the text for the downstream task.

However the computational complexity of self-attention layer is $O(N^2)$, where N is the length of the input. In the general task of argumentation mining it is required given many topics and texts to extract the relevant pieces of information, which means that we have to calculate the outputs of the model for each possible pair of argumentation components. Thus we propose a second approach, we call $BERT_{vect}$. We calculate the following vector representations ([;] represents vector concatenation):

$$x1 = BERT([CLS]\ text1\ [SEP]) \qquad (4)$$

$$x2 = BERT([CLS]\ text2\ [SEP]) \qquad (5)$$

$$class = softmax([x1; x2]W + b) \qquad (6)$$

This approach allows precomputation of vector representations of text, which can be later used in a simple linear model. While this approach is faster, it lacks the cross-sentence attention, which was previously available in $BERT_{basic}$ architecture.

Our final model is $BERT_{se}$, based on the architecture proposed in [16]. It is mainly similar to $BERT_{basic}$ model, but it adds a special self-explaining layer at the end. After the argumentation components are passed through BERT and enriched with contextual information via attention mechanism, the self-explaining layer extracts span representations $h(i,j)$ in the following way:

$$h(i,j) = tanh[W(h_i, h_j, h_i - h_j, h_i \otimes h_j)] \tag{7}$$

$$o(i,j) = \hat{h}^T h(i,j) \tag{8}$$

$$\alpha(i,j) = \frac{\exp o(i,j)}{\sum \exp(o(i,j))} \tag{9}$$

$$\tilde{h} = \sum \alpha(i,j)h(i,j) \tag{10}$$

$W, \hat{h}$ represent learnable parameters, h_i, h_j represent the output on the last layer of BERT corresponding to the i-th and the j-th token. The final prediction is made on the $\tilde{h}$, which gathers the most important span information, instead of the [CLS] token used in $BERT_{basic}$.

We also follow the modification of loss proposed in the original work:

$$Loss = \log p(y|x) + \lambda \sum_{i,j} \alpha^2(i,j) \tag{11}$$

The increase of the λ factor makes the span α distribution sharper, which in turn makes model focus more on a single span.

Additionally we test pretaining on SNLI (Stanford Natural Language Inference corpus) [4] corpora as a strategy for improving quality on argumentation mining subtasks. The corpora consists of pairs of text and hypothesis and the task is to classify whether the hypothesis contradicts, entails or bears no relevance to the text.

While SNLI mostly contains very simple premises which do not require external knowledge, we believe that its properties may improve model behaviour on both the stance classification and relevance detection as it will help model to better extract features required for general purpose natural language understanding. The SNLI corpora is very rich in paraphrases which may help to better process lexically distant pairs of argumentation components.

4 Datasets

In this work we explore the datasets ArgsEN and EviEN introduced in [19]. The ArgsEN corpora introduces two tasks: stance classification and argument quality assessment. In this work we only use the task of stance classification, which consists of 30497 pairs of topic and claim (Fig. 1).

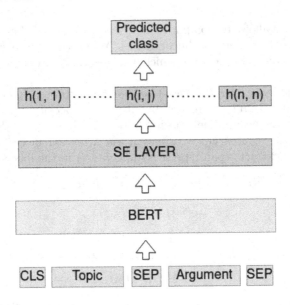

Fig. 1. $BERT_{SE}$ model architecture.

```
Topic: We should end the use of economic sanctions
Claim: Economic sanctions causes safer trading of goods
Stance: PRO

Topic: We should ban algorithmic trading
Claim: algorithmic trading can do a job that humans
are not able to do
Stance: CON
```

EviEN has two tasks: stance classification and evidence detection (relevance detection). The first task is similiar to ArgsEN - given a pair of topic and evidence determine, whether evidence supports or contests the topic. The goal of relevance detection is to determine, whether the evidence is relevant to the topic or not. Combined with the task of stance classification it constitutes for a complete argumentation structure prediction pipeline. The EviEN corpora consists of 35211 samples, all of which are marked up for relevance and 10381 marked for stance.

```
Topic: We should ban the sale of violent video games to minors
Evidence: The storylines of games in this subgenre typically
have strong themes of crime and violence.
Relevance: Relevant
Stance: PRO

Topic: We should legalize doping in sport
Evidence: Contador signed a commitment in which he stated:
```

```
"I am not involved in the Puerto affair nor in any other
doping case"
Relevance: Not relevant
Stance: Not applicable
```

The train-test split in ArgsEN and EviEN is proposed by the authors in such a way, that training data has no topic overlap with the test.

We also explore the transferability of the natural language inference on the tasks of relevance detection and stance classification. As stated before, these problems are similar in formal definition: both tasks are formulated as classification of pairs of statements have similar classes. We use the Stanford NLI [4] to test, whether pretraining the task of textual entailment can improve the quality of argumentation mining models.

5 Experiments

We are using the english weights of $BERT_{base}$ model in all our experiments. Our training hyperparameters also remain the same across all experiments: the models were finetuned for 5 epochs with the learning rate of $2e-5$ on effective batch size of 32. The computations were made on a single Nvidia Tesla V100 GPU using gradient accumulation for $BERT_{SE}$ architecture with the parameter λ of 1.

Table 1. Model metrics on test splits of different subtasks. The values in the "ArgsEN stance" column are the macro-F1 and the accuracy metrics. The values in the "EviEN combined" columns are accuracy metrics for a combined model and a pipeline of models.

Model	ArgsEN stance	EviEN relevance	EviEN stance	EviEN combined
$IBMBERT_{en}$	89.3	–	–	–
$IBMBERT_{17L}$	91.5	–	–	–
$BERT_{basic}$	89.9 (90.)	69.1	86.8	64.7 (52.5)
$BERT_{vect}$	74.5 (74.5)	64.7	72.1	52.3 (46.3)
$BERT_{SE}$	90.0 (90.1)	71.6	87.6	**67.4 (56.1)**
$SNLI_{basic+ZS}$	56.2 (59.0)	45.6	50.4	30.8
$SNLI_{SE+ZS}$	62.3 (62.5)	46.5	57.1	32.3
$SNLI_{basic+FT}$	89.0 (89.0)	**71.9**	88.4	63.3
$SNLI_{SE+FT}$	**90.6 (90.6)**	69.1	**88.8**	64.3

As stated previously the task of stance classification is, given a topic and a relevant text, to classify whether the text supports the topic or contradicts it. We use the ArgsEN dataset, introduced in [19]. We also adapt the metric in the original work, which is macro-F1 averaged by the two classes of PRO/CON. Additionally we calculate accuracy on the whole test split of the dataset.

The authors of [19] provide several models based on the weights of BERT-multilingual model. Those models have IBM prefix in the Table 1. The model $IBMBERT_{17L}$ was trained on a multilingual stance classification corpora, which greatly improved its performance on the english version of the corpora making it currently the best performing model. However, we train our models only on the english data, so our models should be compared to the $IBMBERT_{en}$ model, which was also trained on monolingual corpora.

Models $BERT_{basic}$, $BERT_{SE}$ show performance improvement compared to the models in original work, which may be attributed to the english version of the BERT weights. $BERT_{SE}$ model does not significantly increase model quality over $BERT_{basic}$, while $BERT_{vect}$ displays a relatively low metric.

We train our best performing architectures $BERT_{basic}$ and $BERT_{SE}$ on the SNLI corpora [4] and explore the capabilities of zero-shot approach. The models using zero-shot technique have suffix ZS in their name in the Table 1. For the tasks of stance classification we excluded the "neutral" or "no relation" class from the models prediction. This approach yields negative results as the metrics do not significantly exceed random guess. We additionally explore finetuning after the SNLI pretraining, which greatly improves upon models trained solely on the stance classification task. Such models have FT suffix in the Table 1. The finetuning procedure uses the same hyperparameters as the training, but introduces a newly initialized final linear layer.

We also conduct the same experiments on the stance classification task on the EviEN corpora. The experiments yield similiar results - relative model performance remains identical across the two datasets.

We conduct the same experiments on the tasks of evidence detection on the EviEN corpora as on the tasks of stance classification. The authors of the datasets [19] provide no baselines for the task.

Our results align with the experiments on the stance classification, although the quality of relevance detection is generally much lower. During SNLI pretraining and finetuning we combine the classes of "entailment" and "contradiction", because both those classes imply a relation between the argumentation components.

Finally we explore, how our models perform on a full argumentation mining pipeline. We call this "EviEN combined" task - the goal is, given a topic and a possible evidence text, to classify if it is relevant to the text and further classify its stance if it is indeed relevant.

Our first approach consists of training a model on the combined labels of the EviEN corpora, so the models directly classifies the textual pairs relations as "not relevant", "entailment" and "contradiction". Our second approach consists of pipelining previously trained models: firstly we predict the relevance and only for relevant pairs we predict the stance. As the table shows, combined pipeline improves overall performance on the argumentation mining task.

Despite this setup being the closest to SNLI, as no classes were changed or ignored, because there is a direct mapping between the classes of textual entailment and the 3 classes used in the full argumentation mining pipeline, this approach did not improve the quality of the combined task.

Overall, the experiment results are:

- The $BERT_{basic}$ and $BERT_{SE}$ architectures provide strong baselines for the subtasks of argumentation mining with the latter generally outperforming other models by a margin of several points.
- $BERT_{vect}$ model quality indicates, that the mechanism of cross-sentence attention is highly beneficial to the metrics in the tasks of argumentation mining, when using BERT weights. However it may still be possible to achieve higher computational speed without significant quality loss by using specially trained sentence embeddings, such as [13].
- Zero shot approach based on SNLI pretraining yields no result, as the quality does not noticeably exceed random guess, but SNLI pretraining with further finetuning results in a noticeable improvement over models trained exclusively on the tasks of stance classification and relevance detection.
- The combined approach, trained simultaneously on the tasks of stance classification and evidence detection strongly outperforms pipeline of two models trained on singular tasks.

6 Conclusion

In this work we propose several strong baselines for the subtasks of argumentation mining: stance classification and relevance detection. We find out that cross-sentence attention mechanism is required for a successful model approach using the BERT architecture. Our strongest performing model is $BERT_{SE}$, which was adapted from the natural language inference task and improves upon $BERT_{basic}$ architecture by amplifying cross-sentence attention mechanism with an additional interpretation layer.

We also explore the possibility of combining the subtasks of argumentation mining into classification problem with 3 classes. Models trained using this approach outperform a pipeline of two models for two subtasks on the EviEN dataset.

We also explore the approach of transferring knowledge from SNLI corpora to the ArgsEN and EviEN datasets, which improves model quality and achieves the highest metric on the task of stance classification among the models trained on english data. Additionally we try this technique in a zero shot scenario, but it yields poor results.

We believe that future work might address the problems of exploration of the model's inner knowledge as well as the problem of ways to directly supply model with external information about the world.

Acknowledgements. The study is supported by Russian Science Foundation, project 21-71-30003.

References

1. Aharoni, E., et al.: A benchmark dataset for automatic detection of claims and evidence in the context of controversial topics. In: Proceedings of the First Workshop on Argumentation Mining, pp. 64–68 (2014)
2. Al-Khatib, K., Wachsmuth, H., Hagen, M., Stein, B., Köhler, J.: Webis-debate-16. Zenodo, January 2016. https://doi.org/10.5281/zenodo.3251804
3. Barrón-Cedeño, A., et al.: Overview of CheckThat! 2020: automatic identification and verification of claims in social media. In: Arampatzis, A., et al. (eds.) CLEF 2020. LNCS, vol. 12260, pp. 215–236. Springer, Cham (2020). https://doi.org/10.1007/978-3-030-58219-7_17
4. Bowman, S.R., Angeli, G., Potts, C., Manning, C.D.: A large annotated corpus for learning natural language inference. In: Proceedings of the 2015 Conference on Empirical Methods in Natural Language Processing (EMNLP). Association for Computational Linguistics (2015)
5. Chernodub, A., et al.: TARGER: neural argument mining at your fingertips. In: Proceedings of the 57th Annual Meeting of the Association for Computational Linguistics: System Demonstrations, pp. 195–200 (2019)
6. Cui, L., Cheng, S., Wu, Y., Zhang, Y.: Does BERT solve commonsense task via commonsense knowledge? arXiv preprint arXiv:2008.03945 (2020)
7. Devlin, J., Chang, M.W., Lee, K., Toutanova, K.: BERT: pre-training of deep bidirectional transformers for language understanding. arXiv preprint arXiv:1810.04805 (2018)
8. Ein-Dor, L., et al.: Corpus wide argument mining-a working solution. In: Proceedings of the AAAI Conference on Artificial Intelligence, vol. 34, pp. 7683–7691 (2020)
9. Gleize, M., et al.: Are you convinced? Choosing the more convincing evidence with a Siamese network. arXiv preprint arXiv:1907.08971 (2019)
10. Gretz, S., et al.: A large-scale dataset for argument quality ranking: construction and analysis. In: Proceedings of the AAAI Conference on Artificial Intelligence, vol. 34, pp. 7805–7813 (2020)
11. Levy, R., Bogin, B., Gretz, S., Aharonov, R., Slonim, N.: Towards an argumentative content search engine using weak supervision. In: Proceedings of the 27th International Conference on Computational Linguistics, pp. 2066–2081 (2018)
12. Lippi, M., Torroni, P.: MARGOT: a web server for argumentation mining. Expert Syst. Appl. **65**, 292–303 (2016)
13. Reimers, N., Gurevych, I.: Sentence-BERT: sentence embeddings using Siamese BERT-networks. arXiv preprint arXiv:1908.10084 (2019)
14. Shnarch, E., et al.: Will it blend? Blending weak and strong labeled data in a neural network for argumentation mining. In: Proceedings of the 56th Annual Meeting of the Association for Computational Linguistics (Volume 2: Short Papers), pp. 599–605 (2018)
15. Stab, C., Miller, T., Gurevych, I.: Cross-topic argument mining from heterogeneous sources using attention-based neural networks. arXiv preprint arXiv:1802.05758 (2018)
16. Sun, Z., et al.: Self-explaining structures improve NLP models. arXiv preprint arXiv:2012.01786 (2020)
17. Thorne, J., Vlachos, A., Christodoulopoulos, C., Mittal, A.: FEVER: a large-scale dataset for fact extraction and verification. arXiv preprint arXiv:1803.05355 (2018)

18. Toledo, A., et al.: Automatic argument quality assessment-new datasets and methods. arXiv preprint arXiv:1909.01007 (2019)
19. Toledo-Ronen, O., Orbach, M., Bilu, Y., Spector, A., Slonim, N.: Multilingual argument mining: datasets and analysis. arXiv preprint arXiv:2010.06432 (2020)
20. Vaswani, A., et al.: Attention is all you need. In: Advances in Neural Information Processing Systems, pp. 5998–6008 (2017)

A System for Information Extraction from Scientific Texts in Russian

Elena Bruches[1,2], Anastasia Mezentseva[2], and Tatiana Batura[1,2(✉)]

[1] A. P. Ershov Institute of Informatics Systems, Siberian Branch,
Russian Academy of Sciences, Novosibirsk, Russia
tbatura@iis.nsk.su
[2] Novosibirsk State University, Novosibirsk, Russia
{e.bruches,a.mezentseva1}@g.nsu.ru

Abstract. In this paper, we present a system for information extraction from scientific texts in the Russian language. The system performs several tasks in an end-to-end manner: term recognition, extraction of relations between terms, and term linking with entities from the knowledge base. These tasks are extremely important for information retrieval, recommendation systems, and classification. The advantage of the implemented methods is that the system does not require a large amount of labeled data, which saves time and effort for data labeling and therefore can be applied in low- and mid-resource settings. The source code is publicly available and can be used for different research purposes.

Keywords: Term extraction · Relation extraction · Entity linking · Knowledge base · Weakly supervised learning

1 Introduction

Due to the rapid growth in the number of publications of scientific articles, more and more works have recently appeared devoted to the analysis of various aspects of scientific texts. For example, the paper [10] describes an interface that makes it easier to read scientific articles by highlighting and linking definitions, variables in formulas, etc. Authors of the work [5] propose one of the approaches to the summarization of scientific texts. Texts of this genre contain valuable information about advanced scientific developments, however, this type of text differs from news texts, texts on social networks, etc., in their structure and content. Therefore, it is especially important to adapt and develop methods and algorithms for processing scientific texts.

Common NLP models require a large amount of training data. However, such amounts of data are unavailable for most languages. For example, as far as we know, there are no publicly available datasets for extracting and linking entities in scientific texts in Russian. That is why we focus on the methods which do not require a large amount of labeled data: for term recognition, a weak supervision method is used; for relation extraction, we apply cross-lingual transfer learning; for entity linking, we also implement a language-agnostic method.

© Springer Nature Switzerland AG 2022
A. Pozanenko et al. (Eds.): DAMDID/RCDL 2021, CCIS 1620, pp. 234–245, 2022.
https://doi.org/10.1007/978-3-031-12285-9_15

This paper consists of an introduction, three sections, and a conclusion. The first section provides an overview of works on the tasks of term recognition, relation extraction, and linking terms with entities from the knowledge base. The next section describes the process of preparing and annotating data. In the last section, we give a more detailed description of the system modules, present the algorithms and results of preliminary experiments.

2 Related Work

Terms Extraction. The goal of term extraction is to automatically extract relevant terms from a given text, where terms are sequences of tokens (usually nouns or noun groups) that define a particular concept from a field of science, technology, art, etc. There are several groups of methods for solving this task. The traditional approach solves this task in two stages: firstly, phrases which can be terms are extracted from the text, and then there is a classification step to decide whether this phrase is a term or not. Such an approach is described in [1,19,27]. It allows control term extraction with hand-crafted rules to solve this problem more precisely. But on the other hand, the full context is rarely considered. Another approach solves this task as a sequence labeling task, for example, it is described in [14]. This group of methods takes in account the context to make use of both syntactic and semantic features. The main disadvantage of such an approach is needing quite a large amount of annotated data, as deep learning architectures are used mainly. Some researchers apply methods for topic modeling to extract terms [2]. The underlying idea is that terms represent concepts related to subtopics of domain-specific texts. So revealing topics in the text collection can improve the quality of automatic term extraction.

Relation Extraction. The relation extraction task requires the detection and classification of semantic relations between a pair of entities within a text or sentence. There are different ways to solve this task. A classical approach is to use methods based on lexico-syntactic patterns [11]. Such methods tend to have high precision and low recall since they require manual labor. To overcome this problem, nowadays different machine learning algorithms are applied [20,24]. The distinctive feature of the method [20] is using a neural network for incorporation of both syntactic indicators and the entire sentences into better relation representations. However, it can be difficult to compile a complete list of such indicators. The method described in [24] utilizes special tokens to mark two entities in the sentence which can be used in pre-trained models. This method gives good results in sentence-level relation extraction, but cannot provide information of all the entities (including multiple target entities) in the document at one time. Since entity extraction and relation classification may benefit from having shared information, models for the joint extraction of entities and relations have recently drawn attention. In the paper [13] the authors propose a model architecture, where spans are detected and then relations between spans are classified in the end-to-end fashion. In the paper [22] the authors describe a method for joint entity recognition, relation extraction and event detection in a multitasking way. This method uses separate local task-specific classifiers in the final layer, which can sometimes lead to errors because of a lack of global constraints.

Entity Linking. The entity linking task is the task of matching an entity mentioned in a text with an entity in a structured knowledge base. Usually, this task is considered as a ranking problem and includes several subtasks. The first stage is a candidate generation for the input entity (term) from a knowledge base. For example, it can be done with string match [4], which is rather easy to perform but doesn't solve the problem that the same proper name may refer to more than one named entity. Another approach is dictionary [18] that helps to use taxonomy in knowledge bases but depends on their completeness. Also prior probability [8] can be used in the generation step. In the second stage, one should get embeddings for the mention with its context and for the entity. Nowadays different deep learning-based algorithms are applied for this purpose such as a self-attention mechanism [16], and pre-trained models [25]. Researchers [9] got mention-context vectors by using combined LSTM encoder and bag-of-mention surfaces. This approach lets model the context across datasets. However, authors didn't include more structured knowledge to make representations semantically richer. Then there is a ranking stage to find the most relevant entity for the input term. In some research works this task is solved as a classification task, using different types of classifiers such as naive Bayes classifier [21], SVM classifier [26], deep neural networks [12].

3   Data Description

It is a bit complicated to find the most suitable dataset for our purposes and in the same time open-sourced. Nevertheless, the corpus of scientific papers in Russian RuSERRC [3] solves this problem to some extent. It contains abstracts of 1.680 scientific papers on information technology in Russian, including 80 manually labeled texts with terms and relations. We added an annotation to this corpus by linking selected terms to entities from Wikidata[1].

This corpus contains annotation not only of terms, but also of nested entities (entities that are inside of other entities), for example: "[self-consistent [electric field]] ([самосогласованное [электрическое поле]])". When annotating for entity linking task, we moved from the "largest" entity to the "smaller" nested ones, i.e. if for the very first level the entity was found in the knowledge base, then the nested entities are not annotated.

We linked terms with entities from Wikidata. They have the unique identifier prefixed with "Q", as opposed to relations that have an identifier prefixed with "P". Also, we did not associate terms with entities of the "Scientific article" type. Each entity was annotated by two assessors. The measure of consistency

---

[1] https://www.wikidata.org.

was calculated as the ratio of the number of entities without conflict in the annotation to the total number of entities in the corpus and amounted to 82.33%.

A total of 3386 terms were annotated in the corpus, 1337 of which were associated with entities in Wikidata. The average length of a linked entity is 1.55 tokens, the minimum length is one token, and the maximum is eight tokens.

4   Full System Architecture

We propose a framework for information extraction from scientific papers in Russian. Figure 1 shows its general architecture.

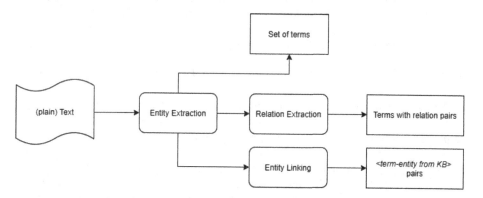

Fig. 1. General architecture of the system.

The system consists of the following modules:

1. The terms extraction module takes a raw text as input and outputs a set of terms from this text;
2. The relation extraction module takes a text and a set of terms obtained from the previous step and outputs a set of term pairs with the specified relation between them;
3. The entity linking module has the same input as the module for relation extraction and outputs terms and the corresponding entities from a knowledge base.

Below we provide a detailed description for each module.

4.1   Entity Recognition

Method Description. During our research, we have not found any comparative large annotated corpora for our purpose. To overcome this problem we decided to use a weak supervision method. The general idea is to train a model on data that were annotated automatically, and then annotate another data with this model, merge these two data sources and train the second model. Thus the algorithm stages are following:

1. Obtaining annotated corpus for the first iteration of model training with a dictionary of terms;
2. Training the model on this corpus;
3. Annotating new texts and the previous ones with the trained model and the dictionary;
4. Training the model on the extended corpus.

The dictionary of terms was obtained in a semi-automated way:

1. We collected 2-, 3-, 4-gramms from the scientific articles, sorted them by TF-IDF value and then manually filtered them;
2. We took titles of articles from Wikipedia, which belong to the category "Science" and then manually filtered them.

In such a way we obtained a dictionary with 17252 terms which is available here[2].

During all stages of our term recognition algorithm, we used the neural network architecture adopted from the BertforTokenClassification class of the Transformers library [23]. Word embeddings were generated by BERT model[3].

The model takes a tokenized text (without any preprocessing) as input and outputs the sequence of labels for the corresponding tokens. Here we use three classes: "B-TERM" (for the first token in a term sequence), "I-TERM" (for the inside token in a term sequence) and "O" (for the token which doesn't belong to a term sequence).

Analysing the results, we noticed that most cases of model errors are wrong term bounds detection. To correct it we wrote some heuristics such as removing a preposition from a term if it starts the term sequence, including the next token written in English after the term sequence, etc.

As a baseline for term extraction we used a dictionary-based approach - the same dictionary was used for automatic text annotation for our model. One may conclude that the dictionary-based approach gives the higher precision for partial match but still gives low recall and F1 in general. Also it can be improved by extending the list of the terms.

Results. We tested the final model on the RuSERRC corpus, which was not used during the model training. To evaluate the model quality, the standard classification metrics were used: precision (P), recall (R) and F-measure (F1). Also, we considered two variants of these metrics: exact match and partial match. In exact match, only full sequences are considered to be correct. In partial match, we considered tokens that have a tag in { "B-TERM", "I-TERM"} as a term. The metrics are shown in Table 1.

---

[2] https://github.com/iis-research-team/ner-rc-russian.
[3] https://huggingface.co/bert-base-multilingual-cased.

Table 1. Term extraction metrics

	Exact match			Partial match		
	P	R	F1	P	R	F1
Baseline	0.25	0.17	0.20	**0.82**	0.34	0.48
BERT	**0.40**	**0.31**	**0.35**	0.77	**0.77**	**0.77**

Relatively low metrics are largely due to the difference between the training set and gold standard annotations. As we annotated the training set with a dictionary, there were not any changes in the token sequences, while in practice the term may not include some terms, has abbreviations, reductions, etc. Analysis of partial match metric reveals that the model is able to recognize a term but it is difficult to define the term boundaries. Considering that the task of term boundary detection is quite difficult even for humans, the obtained metrics are thought to be enough for applying this approach for solving other tasks.

4.2 Relation Extraction

Method Description. There is a lack of annotated data for the relation extraction task in Russian as well. It means that the standard training process of neural networks is difficult. To overcome this issue we applied a pre-trained multilingual model. The main idea is to finetune the pre-trained model on the data in high-resource language. Then evaluate this model on data in Russian. The hypothesis is that information from other languages encoded in the model weights helps to make predictions on data in the target language correctly.

Inspired by [24] we used the R-BERT architecture with pre-trained multilingual BERT model (See footnote 3) and finetuned it on SciERC corpus [15] in English, that has information about relations between scientific terms. Since our task is to define not only the type of relation between two terms but also to define whether two terms are connected by any relation or not, we added samples without any relation to our data. To decrease the imbalance between the number of samples in classes, in the train set we added only 50% of randomly chosen pairs of terms without relation with the distance between such terms less than 10 tokens. There were no such limitations for validation and test sets. We extracted relations only within one sentence.

For relation extraction we also implemented a pattern-based approach (baseline). The main idea is to collect lexico-syntactic patterns manually which can be used as markers for the different kinds of relations. Actually, a lot of samples don't have an explicit marker for a particular relation - it can be found based only on words and text semantics, which is impossible to detect with hand-crafted rules.

Results. To evaluate the model, we used RuSERRC dataset as well. Since the sets of relation types in SciERC and RuSERRC corpora are different, we evaluated the model only on intersected relations: COMPARE, HYPONYM-OF, NO-RELATION, PART-OF, USED-FOR. The overall metrics and metrics by relations (given by the model) are shown in Table 2 and Table 3 correspondingly.

Zero metrics for relation COMPARE show that this relation is understood differently in these datasets. The task of relation extraction and classification is one of the most difficult tasks in NLP. Nevertheless, the obtained metrics show that this approach with some improvements can be used to solve this task without an additional set of annotated data on the target language, although we have yet to investigate this issue.

Due to the fact that we could not find an open-source good-labeled Russian dataset for our purposes we compare our results with previously published results obtained on similar datasets in English. For example, the state-of-the-art result achieved on SciERC with the SpERT (using SciBERT) method is 70.33% for NER and 50.84% for relation extraction. However, according to [7] the same method on the general domain dataset ACE, gives 89.28% and 78.84% f-scores respectively. As can be seen the f-score on the scientific dataset in English is significantly worse due to the complexity of the problem itself. Our results may also be related to insufficient data, as Russian is morphologically rich, which additionally complicates the work of the language model. We plan to study this aspect in the future.

Table 2. Relation extraction metrics

	Precision	Recall	F1
Baseline	0.24	0.30	0.23
Model	**0.27**	**0.37**	**0.25**

Table 3. Metrics by relations

Relation	Precision	Recall	F1
COMPARE	0.0	0.0	0.0
HYPONYM-OF	0.14	0.42	0.21
NO-RELATION	0.97	0.63	0.76
PART-OF	0.15	0.15	0.15
USED-FOR	0.09	0.69	0.17

4.3   Entity Linking

Method Description. We have implemented an algorithm for entity linking. As input data, the algorithm is given a sequence or a single token corresponding to the term. Then the stage of candidates generation, the input string undergoes preprocessing, namely lemmatization using Mystem[4] and conversion to lower-case. Moreover, we take a pre-trained fastText model from Deeppavlov[5] and encoded input term by averaged vectors from it. Entities from Wikidata went through the same stages. Next, two main steps are performed: creating an array of candidates for linking, finding the most suitable entity in the resulting set of candidates.

At the stage of generating candidates, the input string and its 1, 2, 3-grams are compared with the name of the entity and its synonyms. If there is a match, then the entity is added to the candidate list. In addition, if there is a "disambiguation page" in the entity's description, this candidate will be removed from the list of candidates.

For ranking candidates, we use the cosine distance between the two vectors and the threshold value for extra. The first vector is averaged for the input mention and its context of n tokens before the term and the same number after, where $n = 5$. The second vector is also averaged for the name, description, and synonyms of the entity in the knowledge base. In order to take into account the number of matching tokens in candidate and mention, the calculated distances were multiplied by the weighting factor which was computed by the formula:

$$weight = \frac{n\_matching}{n\_all}, \qquad (1)$$

where $n\_matching$ is a number of shared tokens in candidate and mention; n_all is a number of all tokens in entity-mention. The result of the algorithm is the Wikidata item identifier for the input reference.

Results. The algorithm was tested on a corpus with annotated scientific terms from the Data Description section. The metrics are shown in Table 3. We used the following metrics for evaluation: accuracy, the average number of candidates, and top-k accuracy.

Accuracy is the ratio of the number of correctly linked terms to the total number of terms. Since we managed to link not all terms in the corpus, it would be more informative to divide this metric into two: *accuracy* and *linked_accuracy*.

Accuracy takes into account all entities, whereas *linked_accuracy* is calculated only on the set of terms for which the entity was found in the knowledge base in the corpus. Thus, *accuracy* is computed using the formula:

$$accuracy = \frac{n\_correct\_entities}{all\_entities}, \qquad (2)$$

where $n\_correct\_entities$ is the number of correctly linked terms; $all\_entities$ is the number of all terms in the corpus.

Then *linked_accuracy* is calculated by the formula:

$$linked\_accuracy = \frac{n\_correct\_linked\_entities}{n\_all\_linked\_entities}, \qquad (3)$$

---

[4] https://yandex.ru/dev/mystem/.
[5] https://deeppavlov.ai.

where $n\_correct\_linked\_entities$ is the number of correctly linked terms among all linked terms; $n\_all\_linked\_entities$ is the overall number of linked terms in the corpus.

Average Number of Candidates. We also split this metric into two: *averaged_candidates* and *linked_averaged_candidates.*

Averaged_candidates is the average number of candidates for all entities:

$$averaged\_candidates = \frac{\sum_{i=1}^{n} |Candidates_i|}{n\_all\_entities}, \tag{4}$$

where $Candidates_i$ is the set of received candidates for an entity i; $n\_all\_entities$ is the number of all terms in the corpus.

Linked_averaged_candidates is the average number of candidates for the set of terms that managed to be linked. Denote $n\_all\_linked\_entities$ as the number of all terms in the corpus that have a link with the entity from Wikidata, and $Linked\_candidates_i$ as the set of generated candidates for the input term i that was linked with Wikidata. Thus, formula for the *linked_averaged_candidates* metric is:

$$linked\_averaged\_candidates = \frac{\sum_{i=1}^{n} |Linked\_candidates_i|}{n\_all\_linked\_entities}. \tag{5}$$

Top-k accuracy is counted only for a set of terms in the corpus that have a relation with an entity from the knowledge base, in our context k is equal to the number of candidates. This metric is calculated using the formula:

$$top\text{-}k\ accuracy = \frac{num\_correct\_sets}{n\_all\_linked\_entities}, \tag{6}$$

where $num\_correct\_sets$ is the number of candidate sets for the terms which are included in the set of $n\_all\_linked\_entities$, containing the true entity (Table 4).

Table 4. Entity linking metrics

Metric	Baseline pipeline	Final pipeline
Accuracy	**0.71**	0.55
Linked_accuracy	0.53	**0.54**
Averaged_candidates	1.95	**10.29**
Linked_averaged_candidates	2.72	**7.38**
Tok-k accuracy	0.68	**0.76**

A relatively high value for *averaged_candidates*, which shows the generation step works rather properly, has a bad impact on accuracy. By the way, the distinction between the value of *top-k accuracy* and *linked_accuracy* is significantly large. This means that in 76% of cases there is the top candidate in the list, but only in 54% of cases ranking works properly and the output entity is relevant.

To compare results we implemented a simple algorithm for this task. It differs from the final version in two main stages. At the generation step the input string is compared with the name of the entity and its synonyms. If there is a match, then the entity is added to the candidate list. For ranking candidates, we use information about the number of links an entity has to other knowledge bases and the number of relationships of this entity with other entities. The hypothesis is that the more an entity is filled with information, the more relevant it is. Thus, the choice of entity for the input term is determined by the following formula:

$$linked\_entity = max(f(ent_1), ..., f(ent_n)), \qquad (7)$$

where n is the number of entities in the set of candidates,

$$f(ent_i) = numL_{ent} + numR_{ent}, \qquad (8)$$

where $numL_{ent}$ is the number of links to other knowledge bases for this entity; $numR_{ent}$ is the number of relationships of this entity with other entities in the knowledge base.

As for metrics, only accuracy for the baseline is higher than for the final version. Probably, it is due to the significantly low value of average_candidates. The more candidates the more difficult to find the most suitable.

As for the results of other researchers in Entity Linking in Russian, we didn't manage to find any system or even dataset that consists of scientific papers. Nevertheless, there is a suitable dataset in English - STEM-ECR [6]. Authors of this dataset evaluate several systems that work well on open domain data. Otherwise, the scores on their dataset are: exact title match heuristic at 37.8% accuracy, and the best is for Babelfy [17] to DBpedia at 52.6%.

In the future, we plan to implement an approach to identify the semantic similarity of entities using a classifier based on the Siamese network. Moreover, alternative names or synonyms will be used for expanding the list of candidates.

5   Conclusions

In this paper, we presented a system for information extraction from scientific texts in Russian. It consists of three modules. The first module recognizes terms, the second one extracts the relations between the terms found in the previous step, and the third one links terms with entities from the knowledge base. As a result, information from the input text is extracted in a structured form. The experiments were carried out with texts from the information technology domain and, presumably, can be easily adapted to other subject areas. However, in order to draw conclusions regarding the transfer to other domains, in the future, we plan to conduct an additional series of experiments.

Our research is publicly available at https://github.com/iis-research-team/terminator. The results of this study can be useful in the development of systems for unstructured data analysis and expert systems in scientific organizations and higher education institutions.

Acknowledgement. The study was funded by RFBR according to the research project 19-07-01134.

References

1. Bilu, Y., Gretz, S., Cohen, E., Slonim, N.: What if we had no Wikipedia? Domain-independent term extraction from a large news corpus. arXiv preprint arXiv:2009.08240 (2020)
2. Bolshakova, E., Loukachevitch, N., Nokel, M.: Topic models can improve domain term extraction. In: Serdyukov, P., et al. (eds.) ECIR 2013. LNCS, vol. 7814, pp. 684–687. Springer, Heidelberg (2013). https://doi.org/10.1007/978-3-642-36973-5_60
3. Bruches, E., Pauls, A., Batura, T., Isachenko, V.: Entity recognition and relation extraction from scientific and technical texts in Russian. In: 2020 Science and Artificial Intelligence Conference (SAI ence), pp. 41–45. IEEE (2020)
4. Bunescu, R., Paşca, M.: Using encyclopedic knowledge for named entity disambiguation. In: 11th Conference of the European Chapter of the Association for Computational Linguistics. Association for Computational Linguistics, Trento (2006)
5. Cachola, I., Lo, K., Cohan, A., Weld, D.S.: TLDR: extreme summarization of scientific documents. In: EMNLP (2020)
6. D'Souza, J., Hoppe, A., Brack, A., Jaradeh, M.Y., Auer, S., Ewerth, R.: The STEM-ECR dataset: grounding scientific entity references in STEM scholarly content to authoritative encyclopedic and lexicographic sources. In: Proceedings of the 12th Language Resources and Evaluation Conference, pp. 2192–2203. European Language Resources Association, Marseille, May 2020. https://aclanthology.org/2020.lrec-1.268
7. Eberts, M., Ulges, A.: Span-based joint entity and relation extraction with transformer pre-training. arXiv preprint arXiv:1909.07755 (2019)
8. Ganea, O.E., Hofmann, T.: Deep joint entity disambiguation with local neural attention. In: Proceedings of the 2017 Conference on Empirical Methods in Natural Language Processing, pp. 2619–2629. Association for Computational Linguistics, Copenhagen, 2017
9. Gupta, N., Singh, S., Roth, D.: Entity linking via joint encoding of types, descriptions, and context. In: Proceedings of the 2017 Conference on Empirical Methods in Natural Language Processing, pp. 2681–2690. Association for Computational Linguistics, Copenhagen (2017)
10. Head, A., et al.: Augmenting scientific papers with just-in-time, position-sensitive definitions of terms and symbols. ArXiv abs/2009.14237 (2020)
11. Hearst, M.A.: Automatic acquisition of hyponyms from large text corpora. In: COLING 1992 Volume 2: The 14th International Conference on Computational Linguistics (1992). https://www.aclweb.org/anthology/C92-2082
12. Huang, H., Heck, L.P., Ji, H.: Leveraging deep neural networks and knowledge graphs for entity disambiguation. ArXiv abs/1504.07678 (2015)
13. Ji, B., et al.: Span-based joint entity and relation extraction with attention-based span-specific and contextual semantic representations. In: Proceedings of the 28th International Conference on Computational Linguistics, pp. 88–99. International Committee on Computational Linguistics, Barcelona, December 2020. https://aclanthology.org/2020.coling-main.8

14. Kucza, M., Niehues, J., Zenkel, T., Waibel, A., Stüker, S.: Term extraction via neural sequence labeling a comparative evaluation of strategies using recurrent neural networks. In: INTERSPEECH, pp. 2072–2076 (2018)
15. Luan, Y., He, L., Ostendorf, M., Hajishirzi, H.: Multi-task identification of entities, relations, and coreference for scientific knowledge graph construction. In: Proceedings of the 2018 Conference on Empirical Methods in Natural Language Processing, pp. 3219–3232. Association for Computational Linguistics, Brussels (2018)
16. Luo, G., Huang, X., Lin, C.Y., Nie, Z.: Joint entity recognition and disambiguation. In: Proceedings of the 2015 Conference on Empirical Methods in Natural Language Processing, pp. 879–888. Association for Computational Linguistics, Lisbon (2015)
17. Moro, A., Raganato, A., Navigli, R.: Entity linking meets word sense disambiguation: a unified approach. Trans. Assoc. Comput. Linguist. 2, 231–244 (2014). https://aclanthology.org/Q14-1019
18. Pershina, M., He, Y., Grishman, R.: Personalized page rank for named entity disambiguation. In: Proceedings of the 2015 Conference of the North American Chapter of the Association for Computational Linguistics: Human Language Technologies, pp. 238–243. Association for Computational Linguistics, Denver (2015)
19. Stanković, R., Krstev, C., Obradović, I., Lazić, B., Trtovac, A.: Rule-based automatic multi-word term extraction and lemmatization. In: Proceedings of the Tenth International Conference on Language Resources and Evaluation (LREC 2016), pp. 507–514 (2016)
20. Tao, Q., Luo, X., Wang, H.: Enhancing relation extraction using syntactic indicators and sentential contexts. In: 2019 IEEE 31st International Conference on Tools with Artificial Intelligence (ICTAI), pp. 1574–1580 (2019)
21. Varma, V., et al.: IIIT Hyderabad at TAC 2009. In: TAC (2009)
22. Wadden, D., Wennberg, U., Luan, Y., Hajishirzi, H.: Entity, relation, and event extraction with contextualized span representations. In: Proceedings of the 2019 Conference on Empirical Methods in Natural Language Processing and the 9th International Joint Conference on Natural Language Processing (EMNLP-IJCNLP), pp. 5784–5789. Association for Computational Linguistics, Hong Kong, November 2019
23. Wolf, T., et al.: Transformers: state-of-the-art natural language processing. In: Proceedings of the 2020 Conference on Empirical Methods in Natural Language Processing: System Demonstrations, pp. 38–45. Association for Computational Linguistics (2020)
24. Wu, S., He, Y.: Enriching pre-trained language model with entity information for relation classification. In: Proceedings of the 28th ACM International Conference on Information and Knowledge Management, CIKM 2019, pp. 2361–2364. Association for Computing Machinery, New York (2019)
25. Yamada, I., Shindo, H., Takeda, H., Takefuji, Y.: Joint learning of the embedding of words and entities for named entity disambiguation. In: Proceedings of The 20th SIGNLL Conference on Computational Natural Language Learning, pp. 250–259. Association for Computational Linguistics, Berlin (2016)
26. Zhang, W., Su, J., Tan, C.L., Wang, W.T.: Entity linking leveraging automatically generated annotation. In: Proceedings of the 23rd International Conference on Computational Linguistics (Coling 2010), pp. 1290–1298. Coling 2010 Organizing Committee, Beijing (2010)
27. Zhang, Z., Gao, J., Ciravegna, F.: SemRe-rank: improving automatic term extraction by incorporating semantic relatedness with personalised PageRank. ACM Trans. Knowl. Discov. Data (TKDD) 12(5), 1–41 (2018)

Improving Neural Abstractive Summarization with Reliable Sentence Sampling

Daniil Chernyshev[✉] and Boris Dobrov

Lomonosov Moscow State University, Moscow, Russia
chdanorbis@yandex.ru

Abstract. State-of-the-art abstractive summarization models are able to produce summaries for various types of sources with quality comparable to human written texts. However, despite the fluency, the generated summaries are often erroneous due to factual inconsistencies caused by neural hallucinations. In this work, we study possible ways of reducing the hallucination rate during abstractive summarization. We compare three different techniques aimed at improving the correctness of the training procedure: control tokens, truncated loss, and dataset cleaning. To control hallucination rate outside of the training, we propose an improved algorithm for summary sampling - reliable sentence sampling. The algorithm utilizes fact precision metrics to sample the most reliable sentences for an abstractive summary. By conducting the human evaluation, we demonstrate the algorithm's efficiency in preserving summary factual consistency.

Keywords: Abstractive summarization · Hallucinations · Controlled generation

1 Introduction

Abstractive summarization is a form of document summarization where the summary is obtained by extracting salient parts and paraphrasing at the same time. Modern neural summarization models are able to create a summary of different levels of abstraction while achieving a high level of fluency and coherence. However, according to Maynez et al. [17], higher abstraction levels lead to increased information distortion in summary caused by text objects known as neural hallucinations.

Neural hallucinations are an integral part of text generation models which gives them the ability to create text objects not found in input data. Due to the focus of most previous works on extractive summarization datasets, the issue of neural hallucination correctness was poorly studied [13]. By studying the behavior of the existing models on one-sentence abstractive summarization dataset Xsum [18], Maynez et al. [17] found that up to 70% of generated summaries contain neural hallucinations and over 90% of hallucinated statements lead to factual inconsistency. Such a high level of factual hallucination raises serious concern about the reliability of existing methods and postpones their widespread practical application.

© Springer Nature Switzerland AG 2022
A. Pozanenko et al. (Eds.): DAMDID/RCDL 2021, CCIS 1620, pp. 246–261, 2022.
https://doi.org/10.1007/978-3-031-12285-9_16

This paper explores possible ways of reducing hallucinations that may disrupt the fact set. To distinguish acceptable context synonyms from actual factual mistakes, we develop a new dataset for Russian news summarization[1] that exploits the concept of news clusters. We compare three existing methods that reduce the rate of factually incorrect hallucinations and study their effect on abstractive summarization model performance. Dataset cleaning proves to be the most efficient of all, increasing target replication precision up to 19% while also improving copying of named entities.

To improve factual consistency during model inference, we propose a new method for summary sampling - reliable sentence sampling (RSS). In contrast to previous works, our method controls the summary generation process on sentence-level instead of just selecting the most correct variant of a complete summary. We propose new metrics to measure summary reliability and demonstrate the efficiency of the new sampling algorithm in reducing the amount of generally incorrect facts. To study correlation with human judgment, we test our most successful version of RSS against human experts.

2   Related Work

2.1   Abstractive Summarization

One of the first works to apply the sequence-to-sequence model to abstractive summarization task was the work of Rush et al. [20] where an attention-based model for headline generation was proposed. Later See et al. [22] adapted a copy-attention mechanism to solve the out-of-vocabulary issue as well as improve model attention to essential text details. With the introduction of pre-trained language models with global context knowledge like BERT [3], the copy mechanism was no longer necessary for the extraction of important fragments. Liu et al. [16] proposed BERTSum model which adapts BERT encoder and uses a simple multilayer Transformer [23] decoder. Despite BERTSum surpassing copy-attention model scores, the full potential of language models was not utilized due to the lack of text generation tasks in the pretraining phase. Lewis et al. [15] solved this issue by employing denoising pretraining in BART language model which allowed it to be fine-tuned directly to various text generation tasks, including abstractive summarization.

2.2   Hallucinations in Summarization

With the first sequence-to-sequence summarization models, it became evident that generated summaries contained more sophisticated errors than just not following target vocabulary. Cao et al. [1] discovered that almost 30% of generated headlines contained factual mistakes. That claim was later supported in the work of Kryscinski et al. [13] where it was shown that models trained on more complex summarization datasets, such as CNN/Daily Mail, had a similar ratio of factually inconsistent summaries.

---

[1] https://github.com/dciresearch/RNC-dataset.

Several hypotheses of factual inconsistency causes in abstractive summarization models were proposed. Kryscinski et al. [13] argued that the problem lies in the lack of constraints in the summarization task and the existence of incorrect examples in popular for the task datasets that were automatically scraped without validation. The effect of noise in the training dataset is demonstrated in the work of Dusek et al. [4] where authors managed to reduce the ratio of inconsistent text by cleaning noisy examples. Maynez et al. [17] agreed with the lack of constraints in the summarization task, however, it was also pointed out that the problem of factual inconsistency is natural for the text generation models due to their main creation tool - hallucinations.

A neural hallucination is an object generated by the model that is not directly related to the input data. Text generation models, such as abstractive summarization models, use hallucinations to paraphrase text. Maynez et al. [17] showed that during abstractive summarization hallucination rate may reach 70% and in the majority of cases generated hallucinations contain extrinsic information which cannot be validated with the source text. Their analysis of extrinsic hallucinations concluded that in 90% of cases new information was factually incorrect.

2.3   Methods of Improving Factual Consistency

Goodrich et al. [7] proposed using a fact set precision metric to measure the quality of generated summary in terms of fact consistency. Falke et al. [5] used a similar fact precision idea to choose the most factually correct variant among generated summaries, however, instead of extracting facts directly, the authors adopted NLI models to measure correctness sentence-wise. Kryscinski et al. [14] improved this approach by proposing a special model which measures fact consistency between a full source document and generated summaries by classifying each token. Later Zhou et al. [25] expanded the classification approach to detect hallucinations directly by fine-tuning pre-trained language models to distinguish distorted data. A more natural approach to measuring consistency was proposed by Wang et al. [24] where, instead of checking text inference, the correctness is determined by comparing answers found in the source and the summary for the set of questions generated for the summary.

Besides reranking techniques, there are methods that embed knowledge graphs into model attention, append facts to model input, unify entailment checking and text generation models, post-edit final summaries to correct errors, or directly optimize fact precision metrics during training using reinforcement learning [11].

3   Obtaining Dataset for Russian News Summarization

In this work, we focus on studying the possible ways of reducing hallucinations in Russian abstractive summarization models. By the date of writing, only one Russian summarization dataset was proposed - Gazeta [9]. This dataset contains 63435 examples with either extractive or abstractive summaries.

However, our inspection of the dataset revealed that not all text-summary pairs are correct for the summarization task. The example of bad pair is shown in Table 1. The article is about the details of a show hosted at Kremlin, while the summary explains the possible reasons for the event as well as gives some national facts. This example would promote frequent hallucinations of the abstractive summarization model since most fragments of the summary cannot be extracted from the text directly and, thus, can only be guessed.

Table 1. Translated example of bad article-summary pair from test partition of Gazeta dataset. Red - parts that cannot be extracted or derived from the source text.

Article
About 11 thousand spectators saw all the best that is in the culture of Buryatia today. The Buryat State Academic Opera and Ballet Theater, the National Circus, the Buryat National Song and Dance Theater "Baikal" performed in the Kremlin, which became the winner of the show "Everybody Dance!" on the TV channel "Russia", as well as other professional and amateur groups of the region. More than 300 artists from one region on the main stage of the country - it looks like a record for Russia

Summary
On February 25 and 26, Sagaalgan, the Eastern New Year, was celebrated in the Kremlin Palace of Congresses. Buryatia is the center of Russian Buddhism and one of the few regions of the country where the New Year is officially celebrated twice.

Link: https://www.gazeta.ru/social/2020/02/28/12980611.shtml

The existence of this kind of example can be explained by the nature of the method used to obtain the data which was proposed with the Newsroom dataset [8]. The method automatically builds a summary for a news article by exploring description HTML metatags. The problem with this approach is that there is no guarantee that the description metatag would contain an actual summary.

In Russian news writing, it is rare to find an exclusive human-written summaries due to a widespread scheme of article composition know as the Inverted pyramid. In this scheme, the information is presented according to relevance to the reader so that they would need to read only the first few sentences to understand the concept. This way the first paragraph composed of 3–4 sentences (or article lead) can act as a summary for the whole text. But the scheme does not guarantee the repetition of previously presented information. Thus, excluding article lead from the main body and using it as a summary and the remainder of the article as the source text may not be possible due to lack of contextual connection.

The description HTML metatags are commonly filled with the news leads in case the author of the news article decides not to fill them. This can be easily detected by calculating ROUGE-L precision between description and article which would yield a result indicating an almost complete text overlap. However, it is possible that meta-tags were filled incorrectly and the article lead was separated from the main body. In this case, the ROUGE metric may yield results equivalent to an extremely paraphrased summary which, on the other hand, are valuable for abstractive summarization tasks. The same problem occurs if the author did not follow the news article writing guidelines and decided to use description metatag to draw the reader's attention with irrelevant details.

In Gazeta dataset, ROUGE score thresholds were used to filter the described cases of bad description tags, but it is evident that it was not sufficient for author intended mistakes. Thus, a more advanced filtering procedure is required to further refine this dataset for our research.

Another problem is that the task performance of the neural network depends on a number of examples presented in the training dataset and state-of-the-art models utilize datasets with more than 200k examples. This implies that, even after cleaning the Gazeta dataset, it would be required to expand it with more examples to achieve results similar to results on English counterparts. The issue could be alleviated with multilingual MLSUM [21] dataset that contains article-summary pairs from another Russian publisher "Moskovskij Komsomolets". However, the total count of relevant pairs is less than 27k while the low-quality pair filtering procedure did not employ context analysis which means that the dataset carries the same flaws.

Table 2. Dataset Statistics.

	Six publisher dataset	Russian news [2]	CNN [10]	Daily mail [10]	NY times
Mean article length (words)	201.06	220.0	760.50	653.33	800.04
Mean summary length (words)	41.16	52.6	45.70	54.65	45.54
Lead-3 ROUGE-1	34.93	34.69	29.15	40.68	31.85
Lead-3 ROUGE-2	16.84	12.21	11.13	18.36	15.86
Lead-3 ROUGE-L	27.90	27.69	25.95	37.25	23.75

Alternatively, we used an altered Newsroom approach. Instead of following unreliable HTML metatags, we utilize two facts: article lead can act as a summary, and the same event may be covered by different news publishers. By using articles within one news cluster, we construct pseudo-summary-source pairs: lead of the article from one publisher as the summary and full article as the source from another. The eligibility of this approach was demonstrated in previous work [2].

We extracted news articles from six publishers for the past 4 years: "Газета.ru" (Gazeta), "РИА" (RIA), "Интерфакс" (Interfax), "Комерсантъ" (Kommersant), "Дождь" (TV Rain), "Lenta.ru". To obtain news clusters, we used DBSCAN and multilingual paraphrase xlm-r model [19] to build text embeddings for article headlines and leads. As cluster centroids and summary sources, we chose articles from "Комерсантъ" since this publisher focuses on analytics and it was expected to have the most informative leads. We filtered out all summary-source pairs which had either ROUGE-$1_{precison} < 30$ or ROUGE-L > 80 to exclude examples with low relevance or high text overlap.

The resulting dataset contains 206 487 pairs. The data is divided into training (80%), development (10%), and test (10%). The statistics are presented in

Table 2. As it can be seen, the ROUGE scores for the Lead-3 solution for the new dataset are better than previously proposed "Lenta-Комерсантъ" dataset [2] and bear more similarity to existing English counterparts. However, the difference in ROUGE-L scores is negligble which suggests the same ratio of completely extractive fragments in summaries and, thus, the same level of abstractness.

4 Improving Quality During Training

Since abstractive summarization models are prone to hallucinations, it is important that the training procedure won't promote this type of error. Maynez et al. [1] demonstrated that the common training setting used in the field is insufficient and requires additional constraints. In this section, we describe three existing approaches to reducing hallucinations during training.

4.1 Control Tokens

In sequence-to-sequence models, the previous token in the first step of the decoding process is usually set as a BOS token (beginning of the sentence). However, instead of using universal BOS for all sentences, we can set multiple variants that would condition token probability distribution.

In the method proposed by Fillipova [6] for the data-to-text task, these variants of BOS token are set to represent desired hallucination level of generated text. For the training set they are determined according to formulae:

$$\text{BOS}(G, S) = BOS_{\text{hal}(G,S)}) \tag{1}$$

$$\text{hal}(G, S) = \left\lfloor \left(1 - \text{ROUGE-1}_{precision}(G, S)\right) \cdot 5 \right\rfloor \tag{2}$$

where G - target document (gold), S - source document, $\text{BOS}(G, S)$ - function that determines the initial decoding token for text-summary pair. This way the model will learn to retain source document language at BOS_0 and, on the contrary, at BOS_5 generate text with the most diverse vocabulary.

4.2 Truncated Loss

At the late stages of the model training, the examples that meet requirements for multiple learned data patterns may alter established rules. If those examples are correct, this may result in overall performance improvement. However, it is very likely that the most divergent cases of those examples are incorrect for the task and do not follow necessary constraints. In abstractive summarization, such an example would be a target summary that contains external information that cannot be derived from the source text only. This kind of example would promote additional hallucinations since unrelated parts could only be guessed using accumulated knowledge.

To tackle the issue of examples that promote hallucinations at late stages of training, Kang et al. [12] proposed a truncated loss. The idea behind this method is to collect loss statistics and determine normal values for loss function.

First, the loss history of size k is accumulated. Then example truncation threshold C is calculated by taking $1 - c$ quantile over the distribution of losses. After initializing the threshold training loss takes the following form:

$$\hat{L}(H, X) = L(H, X) - \sum_{l(H,x)>C} l(H, x) \qquad (3)$$

$$L(H, X) = \sum_{x \in X} l(H, x) \qquad (4)$$

where H - model, X - batch, $l(H, x)$ - pair-wise loss function for examples $x \in X$. Error history is reset after each threshold calculation and the threshold is recalculated every k steps. Thus, all examples with loss higher than truncation threshold value C are ignored.

4.3 Dataset Cleaning

As it was mentioned before, it is possible that some examples in the training set may promote model-specific errors such as hallucinations that break factual consistency. The straightforward solution is to remove these examples or fix parts that are incorrect in terms of task constraints.

For our dataset, we found that the BERTSum model struggled to reproduce multi-sentence citations due to preprocessing trait, that forces to split them with sentence separator token (e.g. SEP). These long citations are indistinguishable from ordinary text fragments from the model viewpoint and may be paraphrased and combined with any text part, which may create factual inconsistency. We also found that some summaries contained additional named entities that are considered obvious for the reader but cannot be extracted directly from the source text. For example, mentioned in the article "coronavirus" is expanded as "coronavirus COVID-19" in the summary, which is incorrect in terms of task constraints if the source article does not contain the named entity "COVID-19" since there are other types of "coronavirus".

To improve model training performance, we remove the following summary fragments:

- Citations that require inserting SEP token (multi-sentence citations)
- Named entity expansions that are not supported by source article.

5   Reliable Sentence Sampling

The existing approaches for summary generation may use different techniques to improve overall text consistency. However, their efficiency is limited by underlying sampling algorithms that have high computation cost per explored text variant. This happens to be a crucial issue for filtering methods based on factual consistency metrics that require reviewing a large number of possible variants. We propose a general improvement to the sampling procedure that allows to apply these methods more frequently as well as reduce the required number of alternatives at each sampling step.

5.1   Summary Sampling Issues

Neural abstractive summarization models use sampling algorithms to generate text from predicted conditional probability distribution. Wide-spread beam search sampling tracks k most probable hypotheses at each step. However, these hypotheses may not have any significant difference and be synonymous with any in the set.

By increasing beam width k we also increase diversity in a set of hypotheses at the cost of increased memory consumption and slower step iteration. Falke et al. [13] exploit this idea to improve the factual consistency of generated summaries by reranking a large number of final hypotheses. The main flaw of this approach is that the number of possible hypotheses increases exponentially with each beam search step and by the end of summary sampling a large fraction of factually correct hypotheses would be filtered out due to lower support from learned training examples.

This problem cannot be solved with higher k values due to hardware limitations. Applying the hypotheses reranking technique at every sampling step is not an alternative either as it has a very high risk of collapsing probability space to a variant unexplored during model training which may lead to incoherent text.

On the other hand, to properly evaluate text factual correctness, it must consist of finished ideas since the fact set may change with each added word. To alleviate the issue, we exploit the fact that text is naturally structured into finished ideas with sentences.

Algorithm 1. Reliable sentence sampling. Input: k – beam width, source – source text, metric – reliability metric function. Output: summary text

```
function RSS(k, source, metric):
  summary,bestState←SamplerInit()
  while summary[-1]≠EOS:
    cands,result=RSS_step(k,source,summary[-1], bestState)
    if length(cands)=1 and cands[0]=result:
      return summary.concat(result)
    bestHyp,bestState←selectByRelScore(cands, metric, source)
  summary.concat(bestHyp)

function RSS_step(k,source, BOS_token, state):
  hyp,states←BeamSearchInit(k, BOS_token, state)
  cands←[]
  while hyp[0][-1]≠EOS:
    scores, states←DecoderStep(hyp, states)
    hyp←Top_K(hyp, scores)
    for h in hyp:
      if h[-1]=SEP:
        cands.append((h, getState(h, states)))
  final_hyp←hyp[0]
  return cands, final_hyp
```

5.2   Algorithm

Instead of building a summary token-to-token, we generate text sentence-to-sentence. The idea is to generate multiple sentences-candidates by saving each complete sentence found in the probable hypotheses set and then expand the most appropriate by resuming generation from the respective decoding state.

Our implementation is the following (Algorithm 1). Denote sentence separator token as SEN. At each step of beam search for each hypothesis in top-k set, save token sequence and decoder state if the last token in the hypothesis is SEN. Repeat beam search iteration until top-1 hypothesis' last token is not an end-of-text token. This hypothesis is saved in case there are no alternatives in saved sequences (sentences). If the list of saved sequences contains less than two sequences and the only element is equal to the top-1 hypothesis of the final beam search step, then add the top-1 hypothesis to the accumulated summary and return it as the resulting summary. Else choose from the list sentence with the highest reliability score and add it to the summary. Restart decoding procedure considering the last token of summary and recorded decoder state for last added sentence.

5.3   Reliability Scores

There are two types of information unreliableness: local and global. In terms of abstractive summarization, local unreliableness can be any fact introduced in a summary that is not supported directly by the source text. However, these facts may be obvious for the reader and, thus, do not break the overall reliability of the text. For example, in news articles, it is common to omit the year in rare events like elections or championships so this information cannot be directly inferred from the text but would not be a factual mistake if encountered in summary.

Global unreliableness on the other hand is always an incorrect fact regardless of the information source. The problem is that proving that fact is globally unreliable requires processing a full set of relevant information sources.

Thus, it may not be possible to separate local fact unreliableness from global, but by filtering out locally unreliable facts we can ensure global reliability of a summary and improve quality.

Local unreliableness can be measured as fact set difference as was proposed by Goodrich et al. [7]. In our work, we consider facts related to named entities or main events. Using this set, we compute two metrics: named entity overlap (NEO) and named entity fact overlap (NEFO). The formulae are following:

$$NEO(A, S) = \frac{|NE(A) \cap NE(S)|}{|NE(A)|} \qquad (5)$$

$$NEFO(A, S) = \frac{|NE\text{-facts}(A) \cap NE\text{-facts}(S)|}{|NE\text{-facts}(A)|} \qquad (6)$$

where A - generated summary, S - source text, $NE(X)$ - named entities, $NE\text{-facts}(X)$ - named entity facts in form of subject-predicate-object.

6   Implementation

For the base of our experiments, we have chosen BERTSumAbs [16] abstractive summarization model. Since we study ways of reducing hallucination rate, the results of our research would extend to any other sequence-to-sequence neural abstractive summarization model. All settings are the same as in the original model, except we use a special version of BERT pretrained on Russian news language [2] and train model for 100k steps.

For truncated loss, we used a history of $k = 10000$ elements and quantile level $1 - c = 0.7$. For named entity detection and relation extraction, we used Space-ru model for spaCy library. Two types of facts were extracted:

- Object<-Action->Subject - extracts details about the main event
- Named_entity(Object) and Object->Modifier – extracts all modifiers related to named entities.

All words in fact triplets were lemmatized to take into account possible paraphrases.

7   Experiments

7.1   Training-Based Methods

Considered methods do not alter the sampling algorithm or interfere with a fact set directly, so ROUGE-1 precision and NEO would be enough to describe changes in terms of text quality.

Table 3. Evaluation results of training-based methods.

Method	ROUGE-1 precision	NEO
Base model	41.24	58.01
Truncated loss	45.22	58.88
Control tokens	43.22	57.16
Dataset cleaning	**49.16**	**61.65**
Dataset cleaning + Truncated loss	48.34	61.05

The evaluation results are presented in Table 3. As it can be seen, the most efficient method is dataset cleaning. It can be explained by the duality of this method effect: it eases the task by removing parts of the text that are hard to replicate and it reduces the average summary length, thus, lowering the overall amount of compared tokens. Interestingly, loss truncation lowers the quality after dataset cleaning despite being the second most efficient method. This indicates that removed long citations were one of the sources of examples that promote hallucinations.

The inferior performance of control tokens may be explained by a class imbalance of hallucination levels: maximal level on the dataset is 3 (a result of ROUGE filtering during dataset building) and the fraction of examples with this hallu-cination level is 42% while the fraction of examples with level 1 and lower is 11% in total. The model just did not accumulate enough experience to gener-ate text at the lowest hallucination levels which, according to NEO, did lead to inconsistencies in named entity replication.

7.2   Reliable Sentence Sampling

Since our method filters out parts of the generated text and alters the resulting fact set, it is important to additionally compare ROUGE-1 recall with target summary as well as compare NEFO with the source text.

Table 4. Reliable sentence sampling (RSS) results on the test set.

Method	ROUGE-1 PRECISION	ROUGE-1 recall	NEO	NEFO	Cluster NEO	Cluster NEFO
Beam search (k = 5)	49.16	42.34	61.65	39.57	65.82	44.41
RSS NEO	38.72	50.16	**70.77**	46.78	**76.25**	52.13
RSS NEFO	41.78	49.04	59.89	**50.42**	65.93	**62.74**
Lead-3	33.12	50.54	–	–	–	–

Table 4 shows the results of the evaluation. As expected, both variants of reliable sentence sampling (RSS) improve their respective metrics. However, in both cases, ROUGE-1 precision is lower than in summaries obtained by the original beam search algorithm while recall is higher. Comparing with simple extraction of the first 3 sentences of the article (Lead-3), we can see a similar pattern which suggests that RSS may promote more extractive behavior of text generation procedure. This indirectly means that the overall rate of hallucination is also reduced. Another interesting fact is that RSS with NEFO lowers the average NEO. This may be explained by contextual synonyms since the metric does not consider them.

To further study the effect of unreliableness reduction, we compared result-ing summaries with concatenated source-articles from the same news clusters (Table 4, Cluster metrics). Combining articles within one news cluster allows us to approximate the global fact set for the event and, thus, measure the proba-bility of encountering globally unreliable facts.

As it can be seen, all metrics have improved. The difference between original beam search and RSS with NEFO in terms of NEO has been eliminated at the cluster level. This confirms the hypothesis of the effect of contextual synonyms and suggests that the difference of 1–2% should be considered negligible. NEFO shows the largest improvement of 12% at cluster level, but its maximal value is still lower than NEO. This can be explained by strict constraints imposed on fact triplets where the difference in any component is an error.

7.3 Human Evaluation

To evaluate the efficiency of our method, we use new metrics that have not been studied thoroughly. Specifically, we do not have any information about normal values and correlation with human judgment.

To address the issue, we decided to use a crowd-sourcing platform Yandex.Toloka to evaluate the efficiency of reliable sentence sampling with NEFO metric. To build tasks we choose 100 generated summaries from the test set that had more than 3 alternative sentences for the first two steps of RSS algorithm and had ROUGE-2 score higher than 15% (to ensure summary coherence).

For each summary, we built a task in the following form: source text and 3 sentences-alternatives found by RSS algorithm. The task presented to the human judge was to choose the most factually correct and most detailed (in case variants seemed to have an equal level of correctness) variant. We specifically instruct people to consider as a factual mistake only information that is not supported by or even contradicts the source text and use all other mistakes only to break ties. The "most detailed" requirement was determined during our preliminary evaluations where it was found that variants that were universally correct for any article of a similar topic were preferred by low-quality human judges as they fulfilled the factual correctness constraint but did not require reading the source text.

Sentences supplied to human judges had one chosen by RSS, one that was included by the original beam search algorithm in the final summary, and one random alternative among top-5 sentences in terms of NEFO metric, but with a different set of factual mistakes. In other words, we asked human judges to use their knowledge to perform a factual correctness ranking procedure in RSS algorithm.

In total, 200 tasks were given (100 texts with 2 sets of sentences each). Each judge was asked to complete 5 tasks and each task was completed by 10 different people. To control the quality, we assess all tasks after the full completion of the task pool. We set a minimum time threshold equal to the average time of completion multiplied by 75% (to take into account the variance) and include into each set of tasks "honeypot", which could be solved incorrectly only by choosing a random answer.

Table 5. Human evaluation results of reliable sentence sampling.

Support (number of judges)	% of tasks	Average NEFO (RSS variant)	Average NEFO difference	Average NEO (RSS variant)
4–5	14.56	55.0	6.55	59.53
6–7	38.83	67.62	16.3	71,53
8–10	46.60	82.15	26.11	57,63

The results are presented in Table 5. First of all, the decision of RSS algorithm received minimal support of 4 people, and the fraction of cases with support lower than 6 is less than 15%. Almost in half of the tasks the RSS variant had 8 to 10 support. If we study NEFO values, we can see that with a higher difference between the variant chosen by RSS algorithm and closest alternative (most frequent human judge variant) and average NEFO values (for RSS variant) more than 80% there is a high probability that the RSS variant is the most correct. And on the contrary, if NEFO is lower than 60% and the difference is negligible then RSS algorithm might choose a suboptimal variant. The positive correlation between the NEFO score of the RSS variant and human judge support observed in the table proves that the algorithm seeks the most factually correct variant and improves the overall factual consistency of generated summary.

After inspecting tasks with the lowest support, we found the main flaw of NEFO metric: the metric ignores the facts that do not distort the main idea of the sentence but may be incorrect from general knowledge. An example of such a task is demonstrated in Table 6.

Table 6. Example of a task with low support.

Original	Translation
Source text	
Водолазы обнаружили в Черном море не фюзеляж самолета Ту-154, а крупные обломки, передает РИА «Новости» со ссылкой на представителя Южного регионального поисково-спасательного отряда МЧС. «Там не фюзеляж, там крупные обломки, на место направлен аппарат «Фалькон» для обследования местности, по результатам обследования будет приниматься решение об их подъеме», — сообщил он.	The divers found in the Black Sea not the fuselage of the Tu-154 aircraft, but large debris, RIA Novosti reports with reference to a representative of the South Regional Search and Rescue Unit of the Ministry of Emergency Situations. "There is no fuselage, there is large debris, a Falcon apparatus has been sent to the site to survey the terrain, and a decision will be made to lift them based on the results of the survey," he said.
Sentence variants	
Водолазы обнаружили в Черном море не фюзеляж потерпевшего крушение военно-транспортного самолета, а крупные обломки.	The divers found in the Black Sea not the fuselage of the crashed military transport aircraft, but large debris.
Водолазы обнаружили на глубине 27 м в Черном море не фюзеляж, а крупные обломки, сообщил представитель Южного регионального поисково-спасательного отряда МЧС.	Divers found not the fuselage, but large debris at a depth of 27 m in the Black Sea, a representative of the South Regional Search and Rescue Unit of the Ministry of Emergency Situations said.
Водолазы обнаружили на глубине 27 м в Черном море не фюзеляж, а крупные обломки самолета Ту-154, сообщил представитель Южного регионального поисково-спасательного отряда МЧС.	Divers found not the fuselage, but large debris of the Tu-154 aircraft at a depth of 27 m in the Black Sea, a representative of the Southern Regional Search and Rescue Unit of the Ministry of Emergency Situations said.

The text is about additional details on found wreckage of the crashed airplane. Two variants contain additional details about the depth where the wreckage was found. This statement cannot be found in the source article and since NEFO considers numbers as named entities these variants are marked as incorrect. Thus, RSS proceeds with the first variant that does not contain any factual mistakes related to named entities. However, the first variant does contain a factual mistake - it classifies the crashed airplane as a cargo plane while in reality, it is a passenger aircraft. 5 out of 10 judges decided that this is a more serious mistake than unobvious details about the depth and chose the third variant. This suggests that NEFO ranking is similar to the model of perception of the average reader: the correctness of emphasized text fragments (e.g., named entities or citations) has the highest importance.

8   Conclusions

In this work, we studied the possible ways of improving summary reliability by reducing the rate of incorrect hallucinations. From existing methods, dataset cleaning proved to be the most efficient. The second efficient method, truncated loss, showed half as much quality gain, while in conjunction with dataset cleaning it lowers quality of the generated summary. We proposed a new method for summary sampling that improves factual consistency by choosing the most reliable sentences. To measure reliability, we proposed two new metrics: NEO and NEFO. We demonstrated the efficiency of reliable sentence sampling in improving the global reliability of resulting summaries as well as proved the correlation of NEFO ranking with human judgment.

Acknowledgments. The study is supported by Russian Science Foundation, project 21-71-30003.

References

1. Cao, Z., Wei, F., Li, W., Li, S.: Faithful to the original: fact aware neural abstractive summarization. In: Thirty-Second AAAI Conference on Artificial Intelligence, pp. 4784–4791 (2018)
2. Chernyshev, D., Dobrov, B.: Abstractive summarization of Russian news learning on quality media. In: van der Aalst, W.M.P., et al. (eds.) AIST 2020. LNCS, vol. 12602, pp. 96–104. Springer, Cham (2021). https://doi.org/10.1007/978-3-030-72610-2_7
3. Devlin, J., Chang, M.W., Lee, K., Toutanova, K.: BERT: pre-training of deep bidirectional transformers for language understanding. In: Proceedings of the 2019 Conference of the North American Chapter of the Association for Computational Linguistics: Human Language Technologies, Volume 1 (Long and Short Papers), pp. 4171–4186 (2019)
4. Dušek, O., Howcroft, D.M., Rieser, V.: Semantic noise matters for neural natural language generation. In: Proceedings of the 12th International Conference on Natural Language Generation, pp. 421–426 (2019)

5. Falke, T., Ribeiro, L.F.R., Utama, P.A., Dagan, I., Gurevych, I.: Ranking generated summaries by correctness: an interesting but challenging application for natural language inference. In: Proceedings of the 57th Annual Meeting of the Association for Computational Linguistics, pp. 2214–2220 (2019)
6. Filippova, K.: Controlled hallucinations: learning to generate faithfully from noisy data. In: Proceedings of the 2020 Conference on Empirical Methods in Natural Language Processing: Findings, EMNLP 2020, Online Event, 16–20 November 2020. Findings of ACL, vol. EMNLP 2020, pp. 864–870 (2020)
7. Goodrich, B., Rao, V., Liu, P.J., Saleh, M.: Assessing the factual accuracy of generated text. In: Proceedings of the 25th ACM SIGKDD International Conference on Knowledge Discovery and Data Mining, pp. 166–175 (2019)
8. Grusky, M., Naaman, M., Artzi, Y.: Newsroom: a dataset of 1.3 million summaries with diverse extractive strategies. In: Proceedings of the 2018 Conference of the North American Chapter of the Association for Computational Linguistics: Human Language Technologies, Volume 1 (Long Papers), pp. 708–719 (2018)
9. Gusev, I.: Dataset for automatic summarization of Russian news. In: Filchenkov, A., Kauttonen, J., Pivovarova, L. (eds.) AINL 2020. CCIS, vol. 1292, pp. 122–134. Springer, Cham (2020). https://doi.org/10.1007/978-3-030-59082-6_9
10. Hermann, K., et al.: Teaching machines to read and comprehend. In: NIPS (2015)
11. Huang, Y., Feng, X., Feng, X., Qin, B.: The factual inconsistency problem in abstractive text summarization: a survey. ArXiv abs/2104.14839 (2021)
12. Kang, D., Hashimoto, T.B.: Improved natural language generation via loss truncation. In: Proceedings of the 58th Annual Meeting of the Association for Computational Linguistics, pp. 718–731. Association for Computational Linguistics (2020)
13. Kryscinski, W., Keskar, N.S., McCann, B., Xiong, C., Socher, R.: Neural text summarization: a critical evaluation. In: Proceedings of the 2019 Conference on Empirical Methods in Natural Language Processing and the 9th International Joint Conference on Natural Language Processing (EMNLP-IJCNLP), pp. 540–551 (2019)
14. Kryscinski, W., McCann, B., Xiong, C., Socher, R.: Evaluating the factual consistency of abstractive text summarization. In: Proceedings of the 2020 Conference on Empirical Methods in Natural Language Processing (EMNLP), pp. 9332–9346 (2020)
15. Lewis, M., et al.: BART: denoising sequence-to-sequence pre-training for natural language generation, translation, and comprehension. In: Proceedings of the 58th Annual Meeting of the Association for Computational Linguistics, pp. 7871–7880 (2020)
16. Liu, Y., Lapata, M.: Text summarization with pretrained encoders. In: Proceedings of the 2019 Conference on Empirical Methods in Natural Language Processing and the 9th International Joint Conference on Natural Language Processing (EMNLP-IJCNLP), pp. 3730–3740 (2019)
17. Maynez, J., Narayan, S., Bohnet, B., McDonald, R.: On faithfulness and factuality in abstractive summarization. In: Proceedings of the 58th Annual Meeting of the Association for Computational Linguistics, pp. 1906–1919 (2020)
18. Narayan, S., Cohen, S.B., Lapata, M.: Don't give me the details, just the summary! topic-aware convolutional neural networks for extreme summarization. In: Proceedings of the 2018 Conference on Empirical Methods in Natural Language Processing, pp. 1797–1807 (2018)
19. Reimers, N., Gurevych, I.: Making monolingual sentence embeddings multilingual using knowledge distillation. In: Proceedings of the 2020 Conference on Empirical Methods in Natural Language Processing (EMNLP), pp. 4512–4525 (2020)

20. Rush, A.M., Chopra, S., Weston, J.: A neural attention model for abstractive sentence summarization. In: Proceedings of the 2015 Conference on Empirical Methods in Natural Language Processing, pp. 379–389 (2015)
21. Scialom, T., Dray, P.A., Lamprier, S., Piwowarski, B., Staiano, J.: MLSUM: the multilingual summarization corpus. In: Proceedings of the 2020 Conference on Empirical Methods in Natural Language Processing (EMNLP), pp. 8051–8067 (2020)
22. See, A., Liu, P., Manning, C.: Get to the point: summarization with pointer-generator networks. In: Proceedings of the 55th Annual Meeting of the Association for Computational Linguistics (Volume 1: Long Papers), pp. 1073–1083, January 2017
23. Vaswani, A., et al.: Attention is all you need. In: Proceedings of the 31st International Conference on Neural Information Processing Systems, pp. 6000–6010 (2017)
24. Wang, A., Cho, K., Lewis, M.: Asking and answering questions to evaluate the factual consistency of summaries. In: Proceedings of the 58th Annual Meeting of the Association for Computational Linguistics, pp. 5008–5020 (2020)
25. Zhou, C., et al.: Detecting hallucinated content in conditional neural sequence generation. In: Findings of the Association for Computational Linguistics: ACL-IJCNLP 2021, pp. 1393–1404 (2021)

Author Index